So... Grady argues that
device than a sou

THE NEW MIDDLE AGES

BONNIE WHEELER, *Series Editor*

The New Middle Ages is a series dedicated to transdisciplinary studies of medieval cultures, with particular emphasis on recuperating women's history and on feminist and gender analyses. This peer-reviewed series includes both scholarly monographs and essay collections.

PUBLISHED BY PALGRAVE:

Women in the Medieval Islamic World: Power, Patronage, and Piety
   edited by Gavin R. G. Hambly

The Ethics of Nature in the Middle Ages: On Boccaccio's Poetaphysics
   by Gregory B. Stone

Presence and Presentation: Women in the Chinese Literati Tradition
   by Sherry J. Mou

The Lost Love Letters of Heloise and Abelard: Perceptions of Dialogue in Twelfth-Century France
   by Constant J. Mews

Understanding Scholastic Thought with Foucault
   by Philipp W. Rosemann

For Her Good Estate: The Life of Elizabeth de Burgh
   by Frances A. Underhill

Constructions of Widowhood and Virginity in the Middle Ages
   edited by Cindy L. Carlson and Angela Jane Weisl

Motherhood and Mothering in Anglo-Saxon England
   by Mary Dockray-Miller

Listening to Heloise: The Voice of a Twelfth-Century Woman
   edited by Bonnie Wheeler

The Postcolonial Middle Ages
   edited by Jeffrey Jerome Cohen

Chaucer's Pardoner and Gender Theory: Bodies of Discourse
   by Robert S. Sturges

Crossing the Bridge: Comparative Essays on Medieval European and Heian Japanese Women Writers
   edited by Barbara Stevenson and Cynthia Ho

Engaging Words: The Culture of Reading in the Later Middle Ages
   by Laurel Amtower

Robes and Honor: The Medieval World of Investiture
   edited by Stewart Gordon

Representing Rape in Medieval and Early Modern Literature
   edited by Elizabeth Robertson and Christine M. Rose

Same Sex Love and Desire Among Women in the Middle Ages
   edited by Francesca Canadé Sautman and Pamela Sheingorn

Sight and Embodiment in the Middle Ages: Ocular Desires
   by Suzannah Biernoff

Listen, Daughter: The Speculum Virginum and the Formation of Religious Women in the Middle Ages
   edited by Constant J. Mews

Science, the Singular, and the Question of Theology
   by Richard A. Lee, Jr.

Gender in Debate from the Early Middle Ages to the Renaissance
   edited by Thelma S. Fenster and Clare A. Lees

Malory's Morte Darthur: *Remaking Arthurian Tradition*
   by Catherine Batt

The Vernacular Spirit: Essays on Medieval Religious Literature
   edited by Renate Blumenfeld-Kosinski, Duncan Robertson, and Nancy Warren

Popular Piety and Art in the Late Middle Ages: Image Worship and Idolatry in England 1350–1500
   by Kathleen Kamerick

REPRESENTING RIGHTEOUS HEATHENS IN LATE MEDIEVAL ENGLAND

Frank Grady

palgrave
macmillan

REPRESENTING RIGHTEOUS HEATHENS IN LATE MEDIEVAL ENGLAND
© Frank Grady, 2005.

First published in 2005 by
PALGRAVE MACMILLAN™
175 Fifth Avenue, New York, N.Y. 10010 and
Houndmills, Basingstoke, Hampshire, England RG21 6XS
Companies and representatives throughout the world.

PALGRAVE MACMILLAN is the global academic imprint of the Palgrave Macmillan division of St. Martin's Press, LLC and of Palgrave Macmillan Ltd. Macmillan® is a registered trademark in the United States, United Kingdom and other countries. Palgrave is a registered trademark in the European Union and other countries.

ISBN 1–4039–6699–0

Library of Congress Cataloging-in-Publication Data

Grady, Frank.
    Representing righteous heathens in late medieval England / Frank Grady.
        p. cm.—(New Middle Ages)
    Includes bibliographical references and index.
    ISBN 1–4039–6699–0
    1. English literature—Middle English, 1100–1500—History and criticism. 2. Christianity and other religions in literature. 3. Alexander, the Great, 356–323 B.C.—Romances—History and criticism.
4. Christianity and literature—England—History—To 1500. 5. Religion and literature—England—History—To 1500. 6. Trajan, Emperor of Rome, 53–117—In literature. 7. Ethics, Medieval, in literature.
8. Paganism in literature. 9. Jews in literature. I. Title. II. New Middle Ages (Palgrave (Firm))

PR275.R4G73 2005
820.9'3822—dc22   2005047206

A catalogue record for this book is available from the British Library.

Design by Newgen Imaging Systems (P) Ltd., Chennai, India.

First edition: November 2005

10 9 8 7 6 5 4 3 2 1

Printed in the United States of America.

For My Parents

CONTENTS

ACKNOWLEDGMENTS

This book began many years ago as a dissertation written under the guidance of an extraordinary committee: Anne Middleton, my intellectual debts to whom are visible on just about every page; Steven Justice, who reads academic prose with an armor-piercing acuteness; and Gerry Caspary, whose immense learning brought a much-needed breadth to what at times threatened to be a very narrow study. In the years since I have benefitted from the intellectual, spiritual, and material assistance of many friends and colleagues, among them Mark Burkholder, Andrew Cole, Ed Craun, Andy Galloway, Gail Gibson, E. Terrence Jones, Barbara Kachur, Chuck Larson, David Lawton, Gayle Margherita, Alastair Minnis, Maura Nolan, Jim Rhodes, Vance Smith, and John Tolan.

Some of the research and writing of this book was supported by grants from the UM-St. Louis Office of Research Administration, University of Missouri System Research Board, and the Graduate Division of the University of California, Berkeley. Portions of what follows have appeared in print before: a few paragraphs of chapter one have appeared previously in "*Piers Plowman, St. Erkenwald*, and the Rule of Exceptional Salvations," *YLS* 6 (1992): 61–86; part of chapter two in "Machomete and *Mandeville's Travels*," in *Medieval Christian Perceptions of Islam*, ed. John Tolan (New York: Garland Publishing, 1994): 271–88; part of chapter three in "Contextualizing *Alexander and Dindimus*," *YLS* 18 (2004): 81–106; and a few sentences in chapter four in "The Boethian Reader of *Troilus and Criseyde*," *Chaucer Review* 33 (1999): 230–51.

My deepest and most abiding debts, for their boundless good humor, patience, encouragement, and love, are owed to Amy Pawl, who has been living with Trajan and company for as long as she has been living with me, and of course to Emma and Spencer, the most virtuous little pagans I know.

INTRODUCTION

THE RULE OF EXCEPTIONAL SALVATIONS

The best-known story of the salvation of a righteous heathen, both in the Middle Ages and in writing about the Middle Ages, is that of the Roman emperor Trajan, supposedly released from the pains of hell several centuries after his death through the prayers of Pope Gregory the Great. Jacobus de Voragine, in his thirteenth-century *Golden Legend*, tells the story this way:

> Once when the Roman emperor Trajan was hurrying off to war with all possible speed, a widow ran up to him in tears and said: "Be good enough, I beg you, to avenge the blood of my son, who was put to death though he was innocent!" Trajan answered that if he came back from the war safe and sound, he would take care of her case. "And if you die in battle," the widow objected, "who then will see that justice is done?" "Whoever rules after me," Trajan replied. "And what good will it do you," the widow argued, "if someone else rights my loss?" "None at all!" the emperor retorted. "Then wouldn't it be better for you," the woman persisted, "to do me justice yourself and receive the reward, than to pass it on to someone else?" Trajan, moved with compassion, got down from his horse and saw to it that the blood of the innocent was avenged. . .
>
> One day many years after that emperor's death, as Gregory was crossing through Trajan's forum, the emperor's kindness came to his mind, and he went to Saint Peter's basilica and lamented the ruler's errors with bitter tears. The voice of God responded from above, "I have granted your petition and spared Trajan eternal punishment; but from now on be extremely careful not to pray for a damned soul!" Furthermore, John of Damascus, in one of his sermons, relates that as Gregory was pouring forth prayers for Trajan, he heard a divine voice coming to him, which said: "I have heard your voice and I grant pardon to Trajan." Of this (as John says in the same sermon) both East and West are witness.
>
> On this subject some have said that Trajan was restored to life, and in this life obtained grace and merited pardon: thus he attained glory and was not

finally committed to hell nor definitively sentenced to eternal punishment. There are others who have said that Trajan's soul was not simply freed from being sentenced to eternal punishment, but that his sentence was suspended for a time, namely, until the day of the Last Judgment. Others have held that Trajan's punishment was assessed to him *sub conditione* as to place and mode of torment, the condition being that sooner or later Gregory would pray that through the grace of Christ there would be some change in place or mode. Still others, among them John the Deacon who compiled this legend, say that Gregory did not pray, but wept, and often the Lord in his mercy grants what a man, however desirous he might be, would not presume to ask for, and that Trajan's soul was not delivered from hell and given a place in heaven, but was simply freed from the tortures of hell. A soul (he says) can be in hell and yet, through God's mercy, not feel its pains. Then there are those who explain that eternal punishment is twofold, consisting first in the pain of sense and second in the pain of loss, i.e., being deprived of the vision of God. Thus Trajan's punishment would have been remitted as to the first pain but retained as to the second.[1]

The *Golden Legend* was enormously popular and survives in about a thousand manuscripts; Middle English scholars will remember that it is cited in *Piers Plowman* as a source (though not necessarily the only source) of information about the story of Trajan's salvation.[2] But as Gordon Whatley has noted in his thoroughgoing survey of the legend, "No two authorities told the story of Pope Gregory the Great and the Roman Emperor Trajan in identical fashion,"[3] and I quote Jacobus's version here not because it is the authoritative account—there is no authoritative account—but because it is an especially comprehensive one. Not only does the *Legenda* include most of the available details, but it also takes note of the diversity of opinion on the exact nature of the miracle at stake, with its careful litany of "Some have said" and "Others have held" and "Still others. . .say." Stylistically, such perspicacity is perfectly reflective of Jacobus's encyclopedic instinct throughout the *Legenda*, a text that is nothing if not comprehensive. But the range of opinion offered in this passage is less a manifestation of the plenitude of meaning on offer in his notoriously multiple etymologies[4] than an expression of a collective uncertainty about Trajan, an uncertainty that is here attached to the facts of the case—salvation? remission? suspended sentence?—but in fact derives from the principle that seems to underlie it. Certainly Gregory intervened on Trajan's behalf in some way—but was that really such a good idea?

    This uncertainty clearly manifests itself in the treatment afforded to Gregory in this episode. In one account, he prays successfully for Trajan but is immediately admonished by God himself not to indulge in any more such transgressive petitions; John the Deacon, on the other hand, claims that

Gregory did not pray but only wept, as if an inarticulate, inchoate desire for the salvation of the (apparently) damned were more theologically palatable than a focused request. And Jacobus goes on to cite yet another account, in which Gregory's punishment seems to increase in proportion to the relief experienced by the object of his prayers:

> We are told, moreover, that the angel also said: "Because you pleaded for a damned person you are given a dual option: either you will endure two days of torment in purgatory, or you will certainly be harassed your whole life long by infirmities and pains and aches." Gregory chose to be stricken throughout his life by pains rather than to endure two days in purgatory, and so he was constantly struggling with fevers or coping with gout or shaking with severe pains or racked with excruciating stomach cramps. Indeed, he wrote as follows in one of his letters: "I am beset with so much gout and so many kinds of pain that my life is to me a most grievous punishment; daily I grow weak with suffering and sigh expectantly for the remedy of death." In the same vein, in another letter: "The pain I bear is mild at times and very severe at others, but never so mild as to go away and never so severe as to kill me. So it happens that though I am dying daily, I am held at bay by death. I am so thoroughly penetrated by noxious humors that living is an ordeal and I look forward eagerly to death, which I think is the only remedy for my sufferings."[5]

Rhetorically Jacobus makes a canny move here; by quoting from Gregory's letters, he authenticates the pope's suffering without explaining it, that is, without acknowledging the paradox of God granting Gregory's request and punishing him for it at the same time. The angel's "because" points toward this problem, but the reference to Gregory's letters points away from it and seeks to establish a new context for our understanding of the episode. Or rather, two new contexts, overlapping one another: the reference is a historiographical gesture that momentarily displaces legendary narration in favor of quotable documentary discourse, at the same time that it invokes a *contemptus mundi* ethos quite common in hagiography—better a lifetime of physical pain than two days in purgatory, two days separation from God.[6]

As Umberto Eco writes, "[T]hose things that we cannot theorize about, we must narrate."[7] In telling not just the story but the *stories* of Gregory and Trajan, Jacobus begins to reveal for us some of the difficulties inherent in the topic of the righteous heathen, which the Middle Ages found at once so compelling and so troubling. On the one hand, his decision to include the tale was made more or less inevitable by its wide circulation in hagiographies, histories, and theological writing, beginning in the eighth century.[8] But on the other he was fully conscious of the fact that the question of salvation outside the church was a volatile issue, as failing to police adequately the boundary separating pagan from Christian and hell from heaven could have profound consequences. At the bottom of this slippery slope lies not only the suspect idea of universal salvation but serious questions about the utility of the sacraments, for example; heterodox speculations along this line would earn the English Benedictine Uhtred de Boldon official censure a century after Jacobus.[9] The topic obviously required delicate handling, a delicacy expressed in the *Legenda* in its refusal to adjudicate between various alternative explanations.

Such adjudication is precisely the goal of the scholastic theology of the high Middle Ages, but when we turn to Jacobus's contemporary Thomas Aquinas, one of many scholastic writers who took up the issue of Trajan's salvation, we find him struggling with the issue in a similar fashion. In addressing the question "Whether works of intercession profit those who are in hell?" Aquinas writes,

> Concerning the incident of Trajan it may be supposed with probability that he was recalled to life at the prayers of the blessed Gregory, and thus obtained the grace whereby he received the pardon of his sins and in consequence was freed from punishment. The same applied to all those who were miraculously raised from the dead, many of whom were evidently idolaters and damned. For we must say likewise of all such persons that they were consigned to hell, not finally, but as was actually due to their own merits according to justice, and that according to higher causes, in view of which it was foreseen that they would be recalled to life, they were to be disposed of otherwise.
>
> Or we may say with some that Trajan's soul was not freed absolutely from the debt of eternal punishment, but that his punishment was suspended for a time, that is, until the judgment day. Nor does it follow that this is the general result of works of intercession, because things happen differently in accordance with the general law from that which is permitted in particular cases and by privilege. Even so the bounds of human affairs differ from those of the signs of the Divine power as Augustine says. (*De Cura pro Mort.* xvi)[10]

The same regulatory impulse that is implicit in the individual story of Gregory's gout is expressed here as a cautionary rule in Aquinas's *distinctio*: "nor does it follow that this is the general result of works of intercession." And we can see that in seeking to set limits on such admirable works Aquinas is responding in the same way to the same paradox that troubles the *Legenda's* account, if framing the problem somewhat more abstractly. He acknowledges explicitly what is implicit in the *Legenda*, that occasionally specific instances of "privilege" arise (and who in the Christian dispensation is better suited to claim such a privilege than a saint like Gregory?). But he does not pursue the specific dimensions of that privilege; rather, like Jacobus turning to the details of Gregory's suffering rather than the reasons for it,

Aquinas simply moves on to the next question: "Whether works of intercession profit those in purgatory?"[11]

What is clear from these two examples is the topic of the salvation of the
heathen induces one to think in categorical terms (cf. Aquinas's "all such
persons") about a subject that by its very nature urges the violation of conventional doctrinal categories.[12] The problem of the salvation of the heathen is thus not just a problem of ethics or eschatology; it is also a problem
of form, in that it demands that unique incidents be reduced to exemplary
instances, expressions of a rule (a rule that, in Aquinas's formulation, is to be
retired from use almost as soon as it is established). But *exceptio probat regu-
lam*, and establishing a rule—any rule—governing exceptional salvations
inevitably produces the possibility that another exceptional case will present
itself for scrutiny, with each additional instance putting pressure on the
original legislation and threatening to expose either the contingency of its
formulation or the arbitrariness of its application. Such experiments in
form may seek to limit meaning—and what is Jacobus's list of diverse opinions if not a record of formal adjustments designed to fix, once and for all,
the significance of Trajan's salvation by establishing the "true" form of the
story—but they inevitably produce new meanings, new stories, new variations
on a theme.

We should not be surprised to find, then, a diversity of medieval theological opinion about the salvation of the heathen, or of different classes of
heathens—not one rule, but several. Aquinas himself is sensitive to the
distinction between Trajan, saved by an extraordinary intercession, and the
Old Testament patriarchs, the *Gentiles*, who he says were saved by means of
implicit faith in a Mediator if not explicit faith in Christ.[13] Fourteenth-
century nominalist theologians—the *moderni*—speculated about persons
outside the Church meriting salvation by their own virtuous behavior,
ex puris naturalibus, applying the Scholastics' commonplace assertion that
"Facientibus quod in se est Deus non denegat gratiam" ("To he who does
what is in him, God will not deny grace") widely enough to provoke a
vigorous reaction from more Augustinian theologians like Thomas
Bradwardine. Scholarly study of the salvation of the heathen has certainly
not failed to appreciate the diversity of opinion that characterized medieval
theologians' approach to the issue, and indeed such work is valuable in part
because it lets us see late-medieval orthodox discourse as anything but the
monolithic dogma it is often imagined to have been.[14] But it is not enough
to conclude that opinion was generally in agreement on the fate of the
righteous heathen, but divided about the details, because such divisions are,
I would argue, a direct expression of the way the topic is structured,
inevitably part of its repeated and paradoxical attempt to categorize what
amounts to a series of category breakdowns.

The consequences of this structuring become even clearer when we turn from Latin to vernacular texts. Speculation about the salvation of pagans was not, in the later Middle Ages, confined to the clerical discourse of professional theologians and hagiographer-bishops; righteous heathen tales play a prominent part in the rise of what Nicholas Watson has called "vernacular theology," migrating alongside other spiritual and devotional topics and genres—confession and the sins, saints' lives, devotional practices, sermons, mystical visions—from Latin religious writing, popular, elite, and pastoral, to vernacular texts of many kinds throughout the fourteenth century. English writers in particular were drawn to virtuous pagan stories.[15] But when such tales are adapted for their new vernacular circumstances, they are subjected to new and evolving formal conditions, and made open to new formal possibilities—as when, to return again to *Piers Plowman*, a story from a legendary of the saints is incorporated into an allegorical dream vision. If the only questions that need addressing are questions about the "details," then criticism attempting to grapple with Trajan's appearance in *Piers Plowman* is obliged only to specify what those details are—what theological mechanism, what doctrinal codicil, what soteriological belief lies behind the poet's representation of this figure. Does he follow Aquinas, or John the Deacon, or the "some" who say otherwise? And indeed, literary criticism devoted to the topic has often fallen into precisely this pattern, imitating rather than analyzing theological and particularly scholastic responses to the issue.[16] But a critical method that does not give equal weight to the curious and paradoxical rhetorical form of a virtuous pagan story constrains our understanding of such vernacular literary representations. Following the theological line alone tends to obscure the degree to which the Trajan episode and other similar stories are employed by writers in the period in ways that are sometimes "theological" in only the most attenuated sense—ways that demonstrate the appeal of the structural paradox itself, precisely because of the formal innovations it makes possible.[17]

It is not so much that righteous heathen stories take on lives of their own once the topic escapes into the vernacular realm; rather, it's that they necessarily take on the life that individual vernacular authors assign to them, according to the demands of their particular fictions. The way to recover these discrete events—the secret lives of virtuous pagans—is by attending to their formal significance, not form as the expression of theme, "form as content," but form as that which bends and shapes thematic content to fit the special rhetorical circumstances of a dream vision or a chivalric romance or a travel narrative.[18] Thinking formally rather than theologically about the vernacular uses of the righteous heathen thus requires us to think backwards, so to speak, about the issues involved. If a theological approach treats an exceptional salvation as a problem to be solved—by what mechanism, whether orthodox, heterodox or heretical, was this pagan figure saved?—formal analysis treats the same scenario as the solution to another set of problems altogether. At issue is not, or not only, the discovery of the doctrinal principle that underwrites a particular episode's eschatological assumptions, but the specification of the formal or organizational puzzle that the inclusion of a virtuous pagan story permits the writer to solve (or to sidestep, or to defer, or to reframe productively), in the same way that a green girdle, or a steerless ship, or a story about Noah's flood can be employed in order to create certain textual effects. In other words, in the vernacular writing of later medieval England, the topic of the virtuous pagan becomes a topos.

That topos—its formal features and the various ways in which it is employed, ca. 1400—is the subject of this book. We can call this motif the "virtuous pagan scene," and it is comprised of several elements that may appear in full or in various combinations. A structural analysis of such a scene as it appears in late Middle English writing would note the typical inclusion of a dialogue between a representative virtuous pagan and a (usually) Christian interlocutor, in which the latter's inquiry into the "lawe" or spiritual beliefs of the former produces an enumeration of the righteous heathen's many virtues. This list is not so much narrated as—to borrow Hayden White's term—narrativized; that is, the pagan is made to speak in what passes for his own voice and to tell his story in his (and it is, as far as I can tell, uniformly "his") own words.[19] Often this checklist of virtues is made part of an ironic contrast with a Christian viciousness that ill befits a people ostensibly living under the Law of Grace. Sometimes the scene concludes with an exceptional salvation, though this is by no means the sole identifying mark of the virtuous pagan episode, and in fact looking beyond such miraculous events—beyond Trajan—permits us to see just how widespread this motif is in the period. Finally, the scene typically offers some kind of historical self-reflection—not merely the recognition that the esteemed pagan and his Christian counterpart belong to different and sometimes historically remote worlds, but the acknowledgement of some connection or continuity between those worlds that is historical (because of the pagan figure's literal or figurative association with the actual past of the Christian's present) as well as moral (because both individuals recognize similar codes of virtue).

The structuring paradoxes we have already observed in the Latin texts of Jacobus de Voragine and Thomas Aquinas continue to express themselves in the vernacular virtuous pagan scene. While more accounts of exceptional salvations tend to make each one a little less exceptional (hence the evolution of a topos), at the same time the power of any individual episode continues to depend on the singular probity and virtue of the heathen at hand.

Virtuous pagans are thus more than a little like the courtly love objects of medieval romance, at once perfectly unique and perfectly exemplary. Of course, we have long known to treat, say, Malory's noble ladies as the focus of aesthetic and ideological investment, and not as examples of simple portraiture; indeed Malory himself—like most late-medieval writers—was acutely conscious of his place in a long tradition of courtly writing, and was very aware of the formal choices made by his predecessors in their representations of courtly lovers. We have as yet no such critical instincts when it comes to virtuous pagans, in spite of their repeated appearance in recreative, literary texts, doubtless due to their affiliation with a serious theological discourse not widely recognized for formal experimentation. But like the Gueneveres and Isoldes of courtly literature (and of course like fetishized objects in general), they too are subject to laws of genre that govern their textual appearances, and they too are objects of libidinal/emotional investment. Gregory, we might remember, does not so much *decide* to intervene for Trajan as find himself *moved* to do so by his memory of the emperor's kindness. Indeed, from the perspective of desire, *le problème du salut des infidèles* is not per se a problem at all; rather, it is the solution to a problem, the problem of what to do about one's deeply felt sense that the kind of virtues manifested by non-Christian peoples is *real* virtue rather than a pagan imitation, that it is deserving of acknowledgment and reward, and that the rules as they are written often make it difficult to supply such reward. It is no surprise, then, to find that speculation about the fate of admired figures who stand outside the Christian dispensation is as old as Christianity itself, as is the hope that they might be saved.[20]

Nor have such hopes entirely vanished. For a brief example of how a formalist consideration of the virtuous pagan scene can help reveal the contingency of its use, the particular anxieties that it is designed to control, and the emotional investment it seeks to repay, let me turn, somewhat counterintuitively, to a modern example of vernacular theology. In October 1990, Pentecostalist televangelist Larry Lea led a "prayer crusade" in San Francisco at the San Francisco Civic Center; timed to coincide with Halloween, the meeting of his followers from around the nation was designed to loosen the Devil's grip on the city (evidently manifested mostly in the region's gay population). Several of the participants were interviewed by local reporters, and the *San Francisco Chronicle/Examiner* later published excerpts. Dianne DeArman, from Oklahoma City, OK, related this story.

> My hairdresser in Oklahoma was gay, and I loved him. I wrote "Jesus Loves You" in lipstick on his mirror. It really irritated him. The Lord said to me, "I love David like my child. But the Devil has something for him." I told David, "Just give your will to a gentle, loving, easy relationship with the

Lord." David said, "OK, I love the way you love him." He went to church, but they didn't want him because he was very effeminate. There is evil on both sides. David told me, "I won't go anymore, Dianne. I can't bear it when they're so mean to me."

David had AIDS. I laid my hands on him and said, "Be healed, in the name of Jesus. But Lord, if he's not going to be healed, I want him to be happy." He died a week later. No one will ever know the emptiness that came into my heart. David lived in sin. I cried, "Lord, he never quit his practices. Why wasn't I able to help David?" The Lord told me, "Dianne, you introduced him to me, and I love him." And I saw David in this beautiful green field, in paradise. And he had total peace on his face.[21]

This story is, among other things, a welter of cliches—and many of those cliches are the formal markers of an almost archetypal virtuous pagan story. There is the esteemed pagan figure excluded from the church (the sadly typecast but deeply beloved gay hairdresser) whose tale is narrativized in such a way as to foreground his place at the threshold of revelation ("OK, I love the way you love him."), an indictment of Christians who sometimes fail to live up to their own ideals ("There is evil on both sides"), the absence of conversion ("I won't go anymore. . ."), a version of baptism *in extremis* ("I laid my hands on him. . ."), and finally the exceptional salvation itself. There is also the unmistakable sense that this salvation is unique, that is, not generalizable into a principle any larger than the teller's own generosity of spirit (and as is typically the case in such tales the teller's inclusivist tendency is seen as consistent with and indeed analogous to God's ready mercy). To put it another way, the story of the gay hairdresser with AIDS and the compassionate Christian woman that Dianne DeArman here narrates seems to be custom-designed to showcase the fact that the narrator is not the "Christian bigot" that she was accused of being by the protestors gathered outside the Civic Center, but rather a compassionate Christian who endorses and insofar as possible helps to enact God's merciful intention that all should be saved.[22] But evidently they can only be saved one at a time: of course the story *is* custom-built to reflect well on the teller, rather than to suggest that, in the words of *Mandeville's Travels*, "wee knowe not whom God loueth ne whom God hateth" and that one should thus be wary about declaring particular individuals or groups categorically excluded from the visible church.[23] Though such a story could plausibly support a much broader inclusivist position—if this hairdresser can be saved without conversion, why not one of the protesting crowd outside the auditorium, or all of them?—it does not, because what is represented in this tale is not authentic theological speculation but the deployment of a formal topos for particular ends.

The degree to which this very modern story resembles typically medieval ones should put us in mind of another, broader analogy at play in the study of this theme. To summarize once again the medieval circumstances: medieval folk looked back at their pagan predecessors, seeing in them traits that their own contemporary culture valued—truth, justice, righteousness, mercy—and in trying to find ways to register their appreciation of those virtues, to commemorate them, ended up constructing an oft-repeated motif in which a pagan figure participated in a dialogue with a Christian one; in this dialogue the virtue of that pagan was anatomized and recuperated, that is, made intelligible to contemporary ideological conditions through its textual memorialization. In this activity they were looking back at the past from a position of ostensible enlightenment, that is, from the perspective of a Christian revelation that the pagan past (generally) did not share and that marked definitively the difference between the pagan "then" and the medieval "now." Thus the affective impulses that drive this sort of thinking—esteem, admiration, a sense of indebtedness[24]—necessarily had to confront the fact of difference, or as Jill Mann has put it "the problem created by any rupture or discontinuity in experience, outlook or belief between writer and reader. . .[C]hronological distance makes such a rupture inevitable, however various the shapes it may take on."[25]

Of course Mann is here referring, in what was originally a Presidential Address to the New Chaucer Society entitled "Chaucer and Atheism," not to the medieval attempt to understand its past but the modern critical attempt to understand the Middle Ages. In her essay the "enlightenment" that divides then from now is not Christian revelation but in a sense its opposite, modern atheism, which we can let stand for all the various disenchantments that separate us from a Western European medieval culture broadly characterized by its embrace of familiar Catholic certainties, whether we take that disenchantment as the result of a Renaissance invention of subjectivity unknown to the monolithic Middle Ages of the popular imagination, or as the fruit of the analytical tools bequeathed to contemporary criticism by nineteenth-century "founders of discursivity," for example, Freud, Marx, and Darwin.[26] Medieval writers looked to a pagan past that was both a part of their own history and theologically separate from it, because in looking back they found much to value; we moderns look to a medieval past whose fundamental alterity we cannot help but recognize, but that appeals to us for reasons analogous to those medieval motives: their history is a part of our history, and their cultural products—from the aesthetic to the political and economic, and including the theological—continue to have value for us, continue to evoke our esteem and sense of indebtedness. A psychoanalytic account of this claim would of course argue for its inevitability—all forms of desire and *jouissance* are fundamentally analogous[27]—but the comparison is, it seems to me, yet more compelling than this broad diagnosis suggests, by virtue of the methodological similarities that connect modern medieval studies with medievals' attempts to understand their virtuous pagan ancestors. Both enterprises are textual and clerical: textual because they trade not only in the interpretation but the production of texts (and as I will show in the chapters that follow, the production of new texts memorializing virtuous pagans plays an important part in discharging the sense of debt that characterized medieval writers' affective relationship to them); clerical, in a somewhat pejorative sense, because both undertakings are the province of a small literate class that tends to write for a fairly narrow audience (because of limitations in literacy and technology in the later Middle Ages, and institutional professionalization and specialization in our later modern age).[28]

Both undertakings are also historicist, or perhaps it would be more accurate to say that there exists a particularly compelling analogy between medieval writing about righteous heathens and the kind of historicist work that has flourished in medieval literary studies in recent decades. The discourse of the righteous heathen is after all a place where "the desire to speak with the dead," in Stephen Greenblatt's memorable phrase, is often rendered quite literally, in speaking pagan corpses and resuscitated Roman emperors who frequently have quite a lot to say for themselves.[29] Moreover, the medieval attempt to supply the dead with what were imagined to be authentic voices of their own is not entirely dissimilar to the great agon of modern historicist criticism, its variously successful attempts to overcome the general problem of alterity and discontinuity described by Mann without appearing to recuperate the past only for contemporary ideological purposes—attempts, that is, to represent the past insofar as possible in its own terms.[30] Finally, as critics and parodists of the New Historicism have made abundantly clear over the last decade, that approach to thinking and writing about the past has developed some characteristic gestures and rhetorical moves, one of which is the study of the anecdote, in which a discrete historical episode—sometimes obscure, sometimes bizarre—is retold and then analyzed for the ways in which it signifies cultural presuppositions or conventions.[31] It is worth noting that the vernacular literature of the righteous heathen often tends to proceed in much the same way, with an anecdotal account of an exceptional salvation or a curious interview between pagan and Christian that is typically offered to the reader not as an exemplary strategy for handling the occasional anomalous encounter but as the grounds for a certain way of thinking about the past—a way of thinking that, once again, at least on the surface purports to let the past speak for itself.

Of course, I don't mean to suggest that the Middle English discourse of the virtuous pagan somehow anticipates the New Historicism, or that it represents the key to all mythologies when it comes to medieval studies generally. Nevertheless the family resemblances are striking, and suggest that the discourse of the virtuous pagan is relevant not only to our desire to understand medieval uses of antiquity properly, but also the use of the past in general, and our critical, pedagogical, and affective engagement with the Middle Ages in particular. Indeed, this coincidence of the critical and the affective aligns the study of the virtuous pagan with recent self-scrutinizing trends in medieval studies that have begun to explore the part played by our own desires—what Watson calls "the unpredictable play of empathy"—in our apprehension and articulation of the objects of study that we call "medieval."[32] What helps us make our unpredictable attachments meaningful, to ourselves and others, is the imposition of form, whether the form of a virtuous pagan story or the form of a scholarly monograph.

This book, then, explores the structural properties and aesthetic flexibility of the "virtuous pagan scene" as it appears in later Middle English texts, though I think that in "reading for form" I have not merely performed belated New Critical analyses of texts that just happened to escape attention the first time around.[33] The chapters that follow take up the medieval use of the "virtuous pagan scene" as an arrangement of textual elements that makes possible certain kinds of argument, or, once again, certain ways of thinking (and feeling) about the past. Across all formal boundaries—in the travel writing attributed to Sir John Mandeville, the *sui generis* dream visions of Langland and Gower, the ostensibly hagiographic *St. Erkenwald*, the Anglo-Latin chronicle-romances of Alexander the Great, the Middle English Trojan and Arthurian romances, in letter collections and verse sermons—English writers adopt the figure of the righteous heathen and use it to assert different claims to textual power, aspiring to the psychological truth of the dream vision, wielding the authority of the chronicle, invoking the documentary power of the letter, borrowing the sanction and sanctity of the saint's life, adopting the monitory stance of the mirror for princes—creating, that is, a new vernacular discourse of the virtuous pagan.

The first chapter, "The Trouble with Trajan," takes up English versions of the Trajan story, particularly as it appears in *Piers Plowman* and *St. Erkenwald*. Modern criticism of both poems often recapitulates the terms of contemporary medieval debate between those who express sympathy for and identify with the virtuous pagan (whose modern descendants are humanist critics of medieval literature) and those who reject their claims to salvation by means of strict doctrinal appeals (the ancestors of twentieth-century patristic critics). That both sets of critics point to Langland's poem as evidence suggests that *Piers Plowman* itself takes a more comprehensive view of the issue, and I argue that Langland experiments with different strategies in different circumstances, while always calling attention both to the intercessory function of poetry and the status of poetic fictions as potentially authorized and performative uses of language. There is thus a fundamental analogy between what a text tells us about virtuous pagans, and what it does to them: texts narrate (by narrativizing) stories of recovery, like Trajan's exceptional salvation, at the same time that they enact them. Moreover, Langland uses Trajan's appearance to test the limits of personification as a heuristic tool in the third vision's exploration of faculty psychology, since at its most interior moment, *Piers Plowman* suddenly turns outward again, to the historical world of the "trewe knyʒ" Trajan.

To an even greater extent than *Piers*, *St. Erkenwald* defines the question of pagan virtue as a problem for the historical as well as for the moral imagination, beginning with the status of its pagan judge, whose reinsertion into the historical consciousness of the curious Londoners—and their fourteenth-century descendants—comprises the main action of the poem. In fact, *Erkenwald* constantly seeks to literalize *Piers Plowman*'s treatment of the righteous heathen and other aspects of its concern for justice and equity in salvation; topics that are merely discussed and events that are simply remembered in *Piers* are literally enacted in *Erkenwald*, played out in often vivid detail in order to explore and extend the conclusions of its great model and predecessor. To put it another way, *St. Erkenwald*'s use of the motif is designed precisely to place the poem in dialogue with *Piers Plowman* and other contemporary treatments; the author of *St. Erkenwald* self-consciously constructs his poem to take its place in an evolving Middle English tradition by paying particular attention to the basic narrative elements of the virtuous pagan scene. *Erkenwald*, rather than being derivative of *Piers*, is in fact a paradigmatic example of the discourse of the righteous heathen.

The second chapter, "Mandeville's 'Gret Meruaylle,' " concerns the representation of pagans in *The Travels of Sir John Mandeville*, in which the boundaries separating Christian and pagan are literally made the subject of exploration. Here the geographical leveling produced by the conceit of travel—all places, with the telling exception of the Earthly Paradise, turn out to be accessible to the determined traveler—produces a doctrinal leveling that renders all virtue, moral or political, pagan or Christian, past or present, equally admirable and serviceable everywhere, and equally likely to be acceptable to the One God. Reversing the usual order of emphasis, however, I argue that the idea of pagan virtue in *Mandeville's Travels* is its premise rather than its conclusion—that is, that an open-minded, undogmatic appreciation for the integrity and devotion of non-Christian peoples is not a logical position forced on the narrator of the *Travels* by his putative worldly experiences, but rather that the principle of pagan virtue precedes and helps to organize his account of those experiences. Pagan virtue is as much a structural principle in *Mandeville's Travels* as it is an ethical concept; it is a device that, used repeatedly, gives the *Travels* an internal logic and an ideological coherence and helps to suture together its two disparate halves, its pilgrim's account of the Holy Land and its traveler's tale of the Far East.

Chapter 3 takes up "The Middle English Alexander" and the way in which the motif of the virtuous pagan allowed authors to mediate between competing traditions—theological, pastoral, and popular—that diverged considerably in their assessment of Alexander's character and career. Though his historical situation obviates questions of conversion and salvation, the language of those themes, familiar from Mandeville and Langland, is often subtly and not-so-subtly present, and English writers in particular extol his ecumenical spirit, his curiosity, and his virtues, demonstrating an ongoing, lively, and continuously evolving interest in Alexander's deeds that bridges multiple textual traditions, flourishes within and across three different languages, and appeals to all of the literate classes of late-medieval England. Alexander's relations with the Jews were a topic of particular interest, in part because the various legends depict an Alexander whose attitude, alternately reverent and scornful, closely mirrored contemporary Christian positions. Moreover, the romance of Alexander offered in its multifarious texts and genres a means of investigating certain problems of the secular life—particularly the aristocratic life—in a context free of the automatic and definitive solutions mandated by a resolutely Christian moral universe, with its largely fixed hierarchy of value and occupations. The author of the alliterative *Alexander and Dindimus*, for example, uses the virtuous pagan scene to stage a debate about the relative ethical claims of chivalric worldliness and monastic asceticism without the default presence of the Christian dispensation to resolve the issue.

Chapter 4, "The Rhetoric of the Righteous Heathen," takes up three aristocratic fictions in which elements of the virtuous pagan scene play a minor thematic role but important structural ones, governing transitions and conclusions in Chaucer's *Troilus and Criseyde*, Gower's *Confessio amantis*, and the Alliterative *Morte Arthure*. Chaucer's engagement with the standard genres of pagan narrative is complex and critical: skeptical from the *House of Fame* onwards about the intercessory and authorizing power of historical writing, he undermines stoic Boethian rhetoric in his longest pagan romances, rejects *de casibus* tragedy in the *Monk's Tale*, and subverts conversion narratives in the *Man of Law's Tale* and Saracen romances in the *Squire's Tale*, while at the same the time generally avoiding intransigent theologizing *in propria voce*. He takes a very Mandevillean attitude, one could say, which is what makes the apotheosis of Troilus at the end of *Troilus and*

Criseyde a particularly surprising event. I argue that the apotheosis helps to solve (for once in Chaucer's work) the formal problem of the ending by reframing the poem's inescapable *contemptus mundi* moral in the generous context of an exceptional salvation—by drawing, that is, on the discourse of the virtuous pagan. The Priamus episode of the Alliterative *Morte Arthure* serves a similar structural purpose, though in the middle rather than at the end of that poem. As Lee Patterson has shown, the benefits that accrue to the Christian world through the conversion of noble pagans (a symbolic connection with the classical world and non-Christian sources of virtue) can sometimes be achieved without the need for the actual event itself— Priamus's "conversion" is a sort of secular exceptional salvation.[34] Ironically, though, the ease with which the universal aristocratic discourse of chivalry leads to conversion for pagan knights does not solve the problems of Christian ones, who must often choose between chivalric action within history and penitential withdrawal from it. Priamus's joust with Sir Gawain, carefully placed between the two other lengthy single encounters in the poem—Arthur's fight with the Giant of Mont St. Michel, and his interview with the pilgrim-knight Sir Cradoke—anchors the poem's survey of chivalric possibilities in a virtuous pagan scene. Finally, the rhetoric of the righteous heathen helps to organize the most difficult transition in Gower's *Confessio*. Alexander's education by Aristotle is the prime example of the pedagogical relationship of ruler and sage on which Gower's idea of court poetry fundamentally depends, and this "well-tutored Alexander" stands at the center of the *Confessio* and Gower's conception of him. At the same time, though, Gower is aware of the several dangers inherent in this late-medieval rhetoric of exemplarity; in moving from the penitential model of book 6 to the *Fürstenspiegel* of book 7, he draws on the structural resources of the virtuous pagan scene to control these anxieties and to manage the intersection of the amatory and the political in his poem.

The conclusion, "Virtuous Pagans and Virtual Jews," returns to a topic touched on in chapter 3, to explore at greater length the inevitable intersection of the discourse of the righteous heathen with the parallel discourses of medieval antisemitism. Texts that tried simultaneously to confront both medieval Christianity's classical past and its Hebraic roots were familiar enough throughout the Middle Ages; Abelard's twelfth-century *Dialogue of a Philosopher with a Jew, and a Christian* is only the best-known of a host of similar dialogues.[35] But virtuous pagan texts that also feature Jews typically make the confrontation a matter of narrative moment rather than philosophical disputation; moreover, in these stories the role of the Christian character tends to be taken over by a representative pagan. At such moments the uncertainties of the righteous heathen question are simultaneously compounded, as the figure of Christianity's "Other" is suddenly doubled, and radically simplified, as Christian and pagan elide in their opposition to the Jew. Whatever spirit of toleration or ecumenicism the virtuous pagan scene makes possible reaches its limit in these texts, which also marks a limit to the analogy I have urged between modern attempts to recover and enjoy an authentic Middle Ages and medieval representations of the righteous heathen.

CHAPTER 1

THE TROUBLE WITH TRAJAN

The *Golden Legend*'s life of Gregory demonstrates how the story of Trajan's salvation is really two stories, the episode of the emperor and the widow and the historically distinct and distant intervention of the pope. The pairing dates from the earliest life of Gregory and recurs in almost every later version of the story, and this ongoing concern for recording a particular, evocative instance of Trajan's righteous dealings rather than some generalized reputation for justice and truth is striking for the way it creates a balance in the tale between Gregory's act and Trajan's, in which the former's act of pity echoes the latter's act of justice.[1] Nancy Vickers, writing about Dante's use of the Trajan story in the *Purgatorio*, describes how Trajan's function is thus doubled: "[O]n one level, the emperor, acting like God, hears and answers the importunate request of the widow; and on another, God, acting like God, hears and answers the importunate request on Gregory's behalf."[2]

I have argued elsewhere that we can see the relationship between the two parts of the legend as more than just an echo, in fact: Gregory's tears and (possible) prayers also serve as the reward described by the widow.[3] Gregory's pity for the emperor, according to the legend, was inspired by a walk through the Forum of Trajan, and in his response to the Forum's monuments, Gregory participates in a Roman tradition; his tears at once symbolize Christian mercy and Roman honor for the departed, a will and a hope that their virtuous examples not be lost. Gregory's Roman piety and Christian pity are superimposed. The memorials erected to ensure Trajan's lasting memory fulfill their pagan purpose through the intervention of a Christian saint; in addition to the hoped-for measure of fame, Trajan through this coincidence also receives the unexpected bounty of salvation. Moreover, the specific pairing of these two moments implies that the relationship between the Christian present and the non-Christian past is properly an affective one: Gregory is moved to tears by his memory of

Trajan, who was himself moved by the tears (and the logic) of the widow. Though later scholastic and post-scholastic writing about Trajan—and indeed much contemporary critical writing about Trajan—tends to rationalize his salvation by focusing on the particular mechanism by which he was saved, the survival of the widow story preserves the essential sense of esteem on which Gregory's connection to Trajan (and the present's connection to the past) is founded. An emotionally moving awareness of the contiguity between Christian and pagan notions of virtue governs fourteenth-century English treatments of the Trajan legend, in which tears flow freely.

Finally, the two halves of Jacobus's version, and the two cultural strains legible in Gregory's act, indicate the extent to which two different textual traditions comprise the legend. The first is historical, biographical, and entirely secular: the widow story describes an incident from the life of Trajan, a Roman emperor of old, that characterizes the emperor as just but that bears no explicit spiritual significance by itself. Indeed, its nonmiraculous, nontypological content makes it stick out oddly in the hagiographical context in which it appears, and its straightforward account of Trajan's virtuous act makes it unlike the anecdotes of sinful nobles or venal monks that typically prompt the spectacular intervention of a saint.[4] That hagiographical model, of course, represents the second textual tradition involved, the explicitly spiritual side of the story aimed in this instance at celebrating Gregory's holiness and piety, which serve in proper saintly fashion to glorify his God. In texts devoted to the life of the saint, the hagiographical mode naturally dominates, surrounding the Trajan incident and using it as a foundation for the further celebration of Gregory's virtues. But to prove that this need not necessarily be the case—to show that the historical and hagiographical traditions are often in a state of some tension, and even conflict—we can turn to a fourteenth-century Middle English text contemporary with *Piers Plowman*, John Trevisa's translation of Higden's *Polychronicon*.

Higden's account of Trajan is motivated by the historiographic concerns of a chronicler; simply put, in the sequence of Roman rulers he comes after Nerva and before Hadrian. Higden nevertheless is keen to characterize Trajan as a particularly just example of his species, attributing to him a secular version of the golden rule: "I wil be suche an emperour to oþer men as y wolde þat þey were to me and þey were emperours," as Trevisa puts it.[5] And the story of the widow is told in full; after hearing her request and its accompanying arguments about profit and reward, "Traianus was meoved by these wordes, and liȝt doun of his hors, and dede þe womman riȝt, and þerfore he was worþy to have an ymage at Rome." Higden is aware of the legend of Trajan's salvation, of course, though he deals with it in a terse sentence: "For so greet riȝtwisnesse it semeþ þat Seint Gregorie wan his soule out of helle."[6] In the *Polychronicon*, then, the story of the widow is treated as a historical fact, worthy of inclusion because of what it reveals about Trajan's character; it shows him to be worthy of a monument, a historicizing observation not made in the *Legenda*.[7] Gregory's intervention is reduced to a brief notice, almost a footnote, though it too is presented as a fact of history rather than hagiology (despite the slight hedge in "it semeþ"—Latin *videtur*—which stands in here for the various disagreements about the story's details that are spelled out in texts like the *Legenda*).

That at least is the way things work in Higden's text. In his translation, however, Trevisa pauses to offer his own opinion on the question of Trajan's salvation: "So it myȝte seme to a man þat were worse þan wood, and out of riȝt bileve."[8] Trevisa's objection is both noteworthy and symptomatic. In the first place it shows that the Trajan question was far from a settled one in the later fourteenth century; though Trevisa's statement is the baldest rejection I know of, his skepticism on the matter of exceptional and extra-ecclesial salvations was shared by a number of writers and thinkers in the second half of the century, ranging from the episcopal and academic (e.g., Bradwardine's *De causa Dei contra Pelagium* of 1349, or the 1368 censure of Uhtred de Boldon's *clara visio* doctrine) to the more popular and vernacular (e.g., Walter Hilton's *Scale of Perfection*, which attacked "deviant views of redemption" in the 1390s).[9] Trevisa's interjection implies his sense of entitlement to address this issue, and though he does not elaborate on his position his remark can certainly be counted as a very condensed expression of Watson's "vernacular theology."

Associating Trevisa's terse statement with this wider field of theological debate lets us see how the Trajan story expresses the same tension in the Middle English *Polychronicon* as it does in the *Legenda Aurea*, though with the emphasis reversed: Higden's text makes the tale primarily an historical episode, not a hagiographical one, while Trevisa's assertion that it is rather an issue of "riȝt bileve" seeks to call attention to the spiritual and soteriological aspects Higden omits. The one stresses Trajan's virtues in his life as a Roman emperor, while the other suggests that such virtues are hardly sufficient for the afterlife. Whatley, in his survey of six centuries of the Trajan legend, describes four kinds of writers who dealt with the story: early hagiographers who included the tale in their lives of Gregory, twelfth-century humanists (e.g., Abelard and John of Salisbury, who offered Trajan as a model of secular virtues), thirteenth-century scholastics interested in the mechanisms by which Trajan and other righteous heathens might be saved, and fourteenth-century "eclectic" writers—Dante, Langland, and Wyclif—whose versions pursued multiple aims.[10] In fact these four categories can usefully be reduced to two, or at least to two impulses, which are expressed in the two inevitably linked but quite distinct halves of the story itself: the desire to praise Trajan, and the desire to explain him.

". . .on Troianus truthe to thenke"

Twentieth-century readings of Trajan's appearance in *Piers Plowman* almost perfectly reproduce the terms of this original dichotomy, while couching it in a coded language of heterodoxy versus orthodoxy. The two poles of the debate were established by mid-century and can be concisely expressed in two quotations from the work of that era. On the liberal or heterodox side was R.W. Chambers, who claimed in work first published in 1923 that "Whatever may have been the case with Dante, there is no doubt how far Long Will 'was conscious of the distinction between his creed and that of the people.' "[11] On the orthodox side was the Vincentian priest T.P. Dunning, who endeavored in 1943 (the year after Chambers's death) "to prove that Langland, in his treatment of the good heathen in passus XI and XII of B, is completely in accord with the common theological teaching of his time."[12] Later criticism has generally fallen into one or the other of these camps, either asserting definitively that Langland's theory of grace was idiosyncratic and heterodox (or at least "semi-Pelagian"), or claiming with equal certitude that Langland's theology was unexceptionably orthodox, largely Augustinian, and completely in line with the claims of high scholasticism.[13] Recent work on the range of medieval opinion on the issue has tended to confirm the impression that Langland is on the more liberal side—to the left of Dante, certainly—and the case for a semi-Pelagian Langland has acquired a certain kind of canonical status through the *Companion to Piers Plowman*.[14] At the same time, though, A.J. Minnis has recently targeted what he calls the " 'semipelagian' lobby" in a critique of the "quest for nominalism" in the work of Langland and Chaucer that includes a long discussion of Trajan.[15] And Robert Adams acknowledges in the *Companion* that "Until the terms 'orthodox' and 'heretical' can be restricted to a purely historical and descriptive function, and until we have a clearer sense of the theological norms of Langland's day, this debate is likely to continue."[16]

Achieving such purity and clarity seems an optimistic goal at best, since reaching it presents not just methodological problems but also epistemological difficulties. At this point, though, we can at least analyze the terms of this ongoing debate, since historicizing these disagreements about the poem's meaning and the poet's theology reveals the way in which they play out some very familiar themes in twentieth-century Anglo-American medieval studies. For example, when Chambers claims that Langland "was conscious of the distinction between his creed and that of the people," what is pretty clearly implied is the degree to which Langland's creed and Chambers's coincide, an impression confirmed when Chambers goes on to observe that Langland "interprets the baptism of the spirit, the *baptismus flaminis*, in the way that a broad-minded Christian would interpret it in the present day."[17] What Chambers sees in Langland is not so much the influence of the *moderni*, those fourteenth-century nominalist theologians whose work may have been known to him, but the beginnings of modernity itself. In this context his frequent use of phrases borrowed from Shelley's *Defence of Poetry* to characterize Langland (like the "creed" line) make immediate sense; explaining Trajan requires him to assimilate Langland into a Romantic model of authorship, in which his poetry, like that of Shelley's Dante, "may be considered as the bridge thrown over the stream of time, which unites the modern and the ancient world."[18] The title of Chambers's last book, in which his final work on Langland appeared, makes his critical position quite clear: *Man's Unconquerable Mind* is devoted to the study of heroic poets, "men of enormous personalities."

For a pointed and illuminating contrast to Chambers we can turn to Minnis's work. In a characteristically learned and lucid essay on the "quest for nominalism" he begins by acknowledging the difficulty of defining the term and expresses his intention to "use the term 'Nominalist' here *improprie* and *secundum communem usum loquendi*, as found in much recent criticism."[19] The Latin phrase (even more than the title) betrays Minnis's real target in the essay, and its relatively modest conclusions—that neither Chaucer nor Langland employs a very specialized nominalist vocabulary, though both were certainly interested in contemporary theological issues that had "hit the headlines"—belie its interest in answering the twentieth-century critics whose bastardized notion of nominalism exaggerates the controversial nature of both the theology and the poetry of the period. The work of a "card-carrying Nominalist (still using that term according to the common *usus loquendi* of recent criticism)," depending on the topic, "may seem far from controversial,"[20] Minnis argues, and his assertion of the general orthodoxy of Robert Holcot et al. not only seeks to recuperate Chaucer and Langland from the charge of radicalism but also indicts the authors of recent criticism for their wishful and imprecise approach to the question. "A clearer sense of the theological norms of Langland's day," as Adams puts it, thus presumably leads to the conclusion that Langland's poetry stands well within these norms.

This is, or ought to be, familiar ground. Chambers, who was the first to characterize Trajan's appearance in doctrinal terms and who can be counted as the father of this debate, reached his conclusions about the issue via the premise that Langland, like all great "heroic" poets, was a figure "in the world, but not of it," a genius who resisted the tyrannical orthodoxies of his day and communicated that humane spirit of resistance to future generations, because that's what genius poets do. Dunning and those who in his footsteps proclaim the orthodoxy of Langland's account of Trajan assume that all medieval writing not explicitly heretical tends to embody and communicate the doctrinal orthodoxies of the moment—that is simply what medieval writing does. The ongoing conflict in Anglo-American medieval studies between Humanists and Exegetes that Lee Patterson has described as characterizing Chaucer studies after mid-century is here prefigured (and continues to be played out) in the study of Langland, almost a decade before Robertson and Huppé's *Piers Plowman and Scriptural Tradition* (1951) made *Piers* central to the establishment of the exegetical method.[21] What the debate over Trajan ultimately reveals are the complementary critical traditions in *Piers* studies that Anne Middleton has described as the "revelatory" and the "hortatory," the one concerned with examining how the poem harmoniously exemplifies moral and spiritual truths, and the other interested in exploring "the immediate and topical relations between the poet and his contemporary milieu." And though Middleton suggests that these two traditions tend to converge (and that "they are now at most contrasting emphases or starting points for inquiry"[22]), in the matter of Trajan at least a substantial *rapprochement* has yet to occur. The one side proclaims its superior sympathy with the poet, Chambers remarking acidly, in reply to Dunning, that "William Langland must have known what was being preached in the streets of London of his own day";[23] the other side promotes its superior understanding of the period, and decries the potential critical solipsism of those who employ important terms *improprie* and *secundum communem usum loquendi*.

Trajan Speaks

I have of course oversimplified the debate between the modern champions of orthodoxy and heterodoxy here. But that such a debate should exist, and that it should appear irresolvable, should not surprise us. The chief success of Whatley's survey of the Trajan legend is its demonstration that almost every medieval writer used Trajan for a slightly different purpose; it would be more unusual if modern discussions of the legend did not participate in the same grinding of axes and goring of oxen as did their medieval avatars. The historical perspective and bibliographical resources of modern criticism do not always guarantee critical self-awareness. And as I have suggested, the legend itself is comprised of two narratives, either one of which can be used as the premise for quite divergent arguments about the tale's significance. Critics have not been entirely unaware of these facts—the diversity of medieval opinion and the multiple recensions of the legend—and both sides have responded by treating Trajan's appearance in *Piers Plowman* as a morphological problem: determine the shape of Langland's version of the legend—which is not the same as discovering his source—and you have potentially unlocked its meaning. But such an approach reveals as much about the critic's premise as it does about Langland's poem (and can in addition lead to some intemperate statements[24]), whether that premise is systematic—that is, based on the assumption that Langland is guided by a source text, or a doctrine embodied in multiple texts—or sympathetic, that is, based on the premise that Langland is less conventionally indebted to his sources than many of his contemporaries, and offers truly original readings of current issues.

Once again it must be acknowledged that the systematic, exegetical approach has contributed notably to the study of the righteous heathen issue in *Piers* by establishing the range of contemporary opinion on the matter, though ironically the result of such research has often been a more exact idea of the degree to which Langland departs innovatively from scholastic orthodoxy—most particularly in his linking of the Trajan question with issues of baptism and the salvation of the heathen generally, a connection not typically made by scholastic writers.[25] On the other hand, what I have called the "sympathetic" approach to the issue has the virtue of foregrounding—even mimicking—the poem's own apparent bias, which especially in B.11 (Trajan's appearance) and B.12 (Imaginatif's account of the *justi*) seems to lean toward the virtuous heathen. Indeed, Imaginatif seems to occupy a place in the poem analogous to the place of critics outside the poem attempting to account for Trajan's salvation and its consequences, one reason perhaps that Imaginatif's speech at B.12.280–95 tends to figure as largely as it does in sympathetic readings of the Trajan issue.[26] As a number of critics have observed, however, B.12 is not the poem's last word on the fate of non-Christian peoples,[27] and it seems clear that in different contexts within the poem, Langland expresses different kinds of sympathy, and articulates different programs of intercession: a celebration of secular memorialization with Trajan, an uncertain but hopeful appreciation of the contiguity between Christian and pagan wisdom in the case of Imaginatif, and the necessity of episcopally motivated conversion in Anima's long speech in passus 15. All of these imperatives are underwritten and embraced by the theological commonplace of God's wish to extend the possibility of salvation to all people, a principle given vivid expression in Christ's speech during the harrowing of hell, as Nicholas Watson has argued.

Before anatomizing these different strategies of intercession, however, it is worth pausing to engage in another kind of formal analysis of Trajan's initial appearance on the poem, one that explores how Langland's virtuous pagan scene differs not narratively but contextually from other versions of the legend—explores, that is, not which facts and motifs Langland chooses to employ, but where and how he deploys them in his poem. In the first

yes!

place it must be noted that in *Piers Plowman* Trajan speaks—speaks of his own salvation, in his own voice, in the present moment.

> "Ye? baw for bokes!" quod oon was broken out of helle.
> "I Troianus, a trewe kny3t, take witnesse at a pope
> How I was ded and dampned to dwellen in pyne
> For an vncristene creature; clerkes wite be soþe
> That al þe clergie vnder crist ne my3te me cracche fro helle,
> But oonliche loue and leautee and my laweful domes.
> Gregorie wiste þis wel, and wilned to my soule
> Sauacion for sooþnesse þat he sei3 in my werkes.
> And for he wepte and wilned þat I were saued
> Graunted me worþ grace þor3 his grete wille.
> Wiþouten bede biddyng his boone was vnderfongen
> And I saued as ye may see, wiþouten syngynge of masses,
> By loue and by lernyng of my lyuynge in truþe;
> Brou3te me fro bitter peyne þer no biddyng my3te."
>   (B.11.141–53)

As I have already noted, allowing righteous heathens their own voices is characteristic of fourteenth-century Middle English writing; it is a strategy that Langland shares with *Mandeville's Travels*, and one he bequeaths to *St. Erkenwald*. It is also a striking departure, not only from the legend's hagiographical recensions but also a sharp contrast to every other medieval retelling—even Dante, who brings Trajan back to life to be properly converted and baptized, doesn't give him any lines. The sudden appearance of a loquacious authority figure is not at all unusual in a dream vision, of course, and as Kathryn Kerby-Fulton has observed such abrupt materializations probably had a special appeal to medieval readers.[28] Trajan's particular history, though, gives particular consequences to his appearance *in propria persona*.

In the first place, we must appreciate Trajan's speaking as a strategic gesture that allows Langland to emphasize the emperor's virtues at the expense of Gregory's intercession on his behalf. Indeed, Trajan's self-portrait emphasizes exactly that general reputation for probity and justice—his "laweful domes" and "lyuynge in truþe"—that the hagiographical version of the legend eschew in favor of the exemplary episode of the widow (which Langland himself omits). Moreover, insofar as we are invited to treat Trajan as one more character in the poem—something of a vexed issue due in part to critical disagreement about the exact length of his speech[29]—it is only natural to assume that he should emphasize his own merits, just as Scripture or Meed might. Recognizing Trajan's self-regard as a trait does not diminish his virtues, but it does helpfully contextualize their emphasis in the poem.[30]

The style of Trajan's speech is another important aspect of Langland's innovation here. For Trajan's appearance to be intelligible to his readers, who will be encountering a familiar figure in a thoroughly unfamiliar context, Langland must make that appearance both thematically relevant and stylistically appealing and appropriate. Thematically of course Trajan's presence furthers the discussion of equity in salvation that had begun at the end of passus 10, when Will advanced the hard cases of Mary Magdalene, St. Paul, and the good thief—a discussion that was itself connected to the themes of mede and mercede from the earliest parts of the poem, topics specifically invoked again in the discussion of Trajan's salvation. Stylistically, Trajan's introductory "Baw for bokes!" marks him as a colloquial speaker (compare the "baw" of Trajan's evil twin, the obstreperous brewer of B.19.400), and his attempt to offer his self-identification as a kind of testimony—"I Troianus, a trewe knyȝt, take witnesse at a pope"—connects him with the practices of English common law rather than hagiographical or theological discourse. In short, Langland brings Trajan into the discursive environment of his poem by "Englishing" him, a gesture that leaves Trajan doubly domesticated: he is a pagan deemed acceptable to God and worthy of salvation, and he is a foreign and even exotic figure made intelligible to the audience of an English vernacular poem.

What makes Trajan exotic in the poem is not just his historical nationality, however; domesticating and "medievalizing" pagan heroes is a common practice, especially in chivalric romance. Rather, it is the very fact of his historical existence that makes him unique: in a poem populated by ahistorical personifications like Conscience and Meed and transhistorical figures like Christ, Trajan is the only "real" historical person to appear. And his singular status makes the location of his appearance all the more unexpected, even odd. Long acquaintance with Trajan's presence in the inner vision of the third dream of the poem has inured us to surprise, but given the essentially cognitive interests of that dream and its dependence on faculty psychology, it cannot be other than strange to find a deceased Roman emperor at the center of its parade of personifications. Both the names of these personifications and the poetic signatures by means of which they are linked to the dreamer (and thus to the poet) lead us to the conclusion that they are, as Joseph Wittig has put it, "those aspects of his personality or those data of his experience which the process of his 'conversion'. . .systematically calls into prominence."[31] But it is as hard to view Trajan as an aspect of the dreamer's personality as it is to think of him as an element in the dreamer's experience; while we might credit the dreamer with having had some kind of textual experience of Trajan—for example, an encounter with the *Legenda sanctorum* cited at B.11.160—that experience would be qualitatively different from the more typical bodily experience evoked in this dream by

figures like *Concupiscencia carnis* and "Couetise of eiȝes," or even the view of
the natural world offered in the mirror of Middle-Earth. In short, even an
indisputably successful search for the source of Langland's acquaintance
with the Trajan legend would still leave unexplained the nature of the
dreamer's relationship to Trajan—whether he comes from inside or outside,
or whether his presence creates more problems than it solves. Just as Trajan
stands in two places culturally, in pagan Rome and in the heaven of the
medieval Christian West, he also simultaneously occupies two places in
the poem, the "inside" of faculty psychology and the "outside" of history.

Trajan's appearance certainly does solve some local problems for
Langland. In the largest perspective, the inner dream in which he plays so
prominent a part seems to have been the key to breaking the impasse at the
end of the A-text, and thus the key to the entire B-continuation.[32] Within
the inner dream itself, Trajan provides a necessary contradiction to Will's
provisional conclusions about baptism and salvation. Immediately before
Trajan's appearance Scripture agrees with Will's aside about the efficacy of
mercy, responding with a quotation from Psalm 144: "Misericordia eius
super omnia opera eius." This is the first time in the third vision that an
authority figure has agreed with Will, as Simpson notes,[33] but mercy is not
the theme of Will's speech; the sufficiency of baptism is, and in one sense
Scripture's quotation is more of an automatic gesture than an authentic
piece of argument, indisputably true but apparently beside the point.
Scripture's quotation can also be read as an oblique way of conjuring up
Trajan, arguably an individual who has specially benefitted from divine
mercy; in any case Trajan certainly returns the poem to the point, contradicting Will by stressing the fact of his salvation despite his failure to be
baptized. His interruption is thus structurally in keeping with the rest of the
vision, and indeed the rest of the poem, which as Middleton has shown
depends for its narrative progress on the animus-filled encounters in which
one figure—usually the dreamer—is sharply contradicted by another.[34]

Trajan's arrival is also thematically appropriate; as Wittig observes, "The
details of the Trajan legend are rhetorically ideal for shattering Will's false
security."[35] The discussion with Scripture in B.11 returns to an earlier
discussion of the sufficiency of baptism and related soteriological issues that
had begun in passus 10 (343ff.), before the inner dream; it even reuses the
analogy of servants falling in debt ("arerage") to their masters that Will
had employed at B.10.469 to distinguish between learned folk—clerks and
reeves who are likely to fall into debt because of their greater responsibility
to "kepen the lordes catel"—and unlearned servants who "selde fallen in
arerage," where avoiding debt is equated with avoiding sin. In passus 11 Will
expands and generalizes this analogy; at B.11.127–36 the servant ("cherl")
stands for all Christians, who may "renne in arerage and rome fro home"

but who can never "reneye" or abjure the Christianity bestowed upon them by baptism. Will enlarges the scope of the analogy here under the threat of Scripture's anecdotal citation of Matthew 22:4, "Many are called but fewe are chosen." He acknowledges the extent of the *multi*, remarking that ". . .crist cleped vs alle, come if we wolde, / Sarȝens and scismatikes and so he dide þe Iewes" (B.11.119–20), and then tries to fix as exactly as possible the number of the *pauci* in order to determine "Wheiþer I were chosen or noȝt chosen" (117). The invocation of "Sarȝens. . .scismatikes and. . . Iewes" does not, at this point, look forward to a universalist approach to salvation; rather it is part of a logical process of winnowing: Saracens et al. are among the *multi* called but historically not the *pauci* chosen, for they have embraced the wrong creed, or failed to adopt the right one, a point Langland will make quite explicitly through Anima in passus 15. Through the exclusion of these non-Christians—an exclusion so obvious that it doesn't even need to be articulated (because the salvation of the heathen is not a live issue yet, not for another twenty lines[36])—a hopeful Will effects a reduction of the *pauci* to baptized Christians who cannot ever be unbaptized.[37]

It is Trajan's job to explode this scheme, by identifying himself both as one of the few who have been saved and as a "Sarsen," specifically invoking the term Will had used a few lines earlier to exclude a whole class or people from salvation. In fact that term points to the importance of Trajan's historicity; only a character both historically real and historically saved can provide an adequate response to Will's implied assumption about the historical exclusion of Saracens, schismatics, and Jews. But if in one sense the historical nature of Trajan's case is "rhetorically ideal" for the circumstances, in another it utterly breaks the fiction of the third vision. Indeed, Trajan's appearance offers a considerable challenge to the use of allegory and personification as tools for the understanding of the self and its psychology; at its most interior moment, *Piers Plowman* suddenly turns outward again, to the real and historical world. And it does so by authorial fiat; just as he is saved contrary to all Will's assumptions about salvation, Trajan appears in the poem contrary to all expectation, in defiance of any possible expectation. He bursts onto the scene like the eponymous Piers, and like Piers he steers the poem away from an allegory on the verge of a nervous breakdown—in passus 5 the aimless pilgrimage, in passus 11 a paralyzing anxiety about salvation that cannot be resolved by individual faculty psychology—toward the real world of work and poverty. Both the injunction that immediately follows Trajan's appearance ("Wel ouȝte ye lordes þat lawes kepe þis lesson haue in mynde / And on Troianus truþe to þenke, and do truþe to þe peple," B.11.159–60) and the long passage in praise of patient poverty that leads up to the mirror of Middle-Earth point the way toward an active life in the world, whether lordly or simple, evoking that world in a way that the interior dialogues of the third vision, for all their talk of DoWel, cannot successfully achieve.

James Simpson has described the process of "deconstruction" that characterizes the third vision in *Piers Plowman*, the way in which, over the course of the dream, both the discourses of satire and academic debate are put into play only to be found wanting. The poem, he argues, engages *in* these discourses precisely so as to engage *with* them, to explore them in order to exhaust them.[38] Trajan certainly has an important role to play in this process, both in the challenge he poses to Will's soteriological smugness and in the jolt his sudden appearance gives to the larger psychological motif of the dream. Either Langland's notion of faculty psychology is capacious enough to include Trajan's contribution as one of its elements, in which case we must conclude that Langland's approach to the allegorical representation of personality differs importantly from the monastic moral psychology usually taken to be his chief source for the structure of this vision, or Langland consciously uses Trajan to fracture the allegorical model that has to this point functioned as the dream's organizing principle. The conclusion is essentially the same in either case: the structural consequence of Trajan's appearance is the exposure of the limitations of the faculty model both in defining individual psychology and in establishing imperatives for moral action in the world.[39]

Paradoxically Trajan also represents an integrative impulse, alongside his deconstructive function. The integration of disparate cultures and *cultus* is rather obviously what an interest in the salvation of non-Christians amounts to, and though Trajan directs his initial remarks toward the topic of baptism, the fact that the issue of extra-ecclesial salvation recurs later in the poem suggests that Will's encounter with Trajan is the place where the dreamer first begins to learn about a subject of notable interest to Langland, and one that—to his mind at least—is best explored in stages. Will's failure to interact with Trajan in passus 11, which is another thing that makes Trajan unique in the third vision, also suggests that his appearance is only a first step. Though Trajan's story is explicitly about a kind of dialogue, an exchange between pagan and Christian culture, there is actually precious little of it that takes place here (however long we judge Trajan's speech to be). In fact a lack of direct cross-cultural interaction is a feature of the Gregory–Trajan legend, too; Trajan may talk to the widow, but he and Gregory do not converse—that scenario has to wait for *St. Erkenwald*. Perhaps for Langland it was innovation enough to get Trajan to speak at all and tell his own story.[40]

The way Trajan tells that story is one means by which Langland expresses this episode's integrative impulse. Through Trajan he selectively emphasizes the details of Gregory's intervention, stressing not so much what Gregory does (though he apparently favors the idea of the pope's tears rather than his prayers), but what he knows and learns about Trajan's virtues: "And I saued as ye may see, wiþouten syngynge of masses, / By loue and by lernyng of my lyuynge in truþe"(B.11.150–51). Gregory's "will" that Trajan be granted grace arises from his appreciation of Trajan's virtues, specifically his "truþe," a word that appears several times in several forms in the passage. The advice for lords "on Troianus truþe to þenke, and do truþe to þe peple" implies that Trajan's "truþe" still has an active role in a morally ordered life; moreover, though Trajan's "truþe" and the "truþe" that the lords can and should do may not be identical, they presumably share a strong family resemblance. Trajan's exemplary status is the corollary Langland draws from the story of his salvation (as did Abelard and John of Salisbury before him), though it is not something that in the hagiographic versions of the legend ever occurs to Gregory. This assumption—that if Trajan was "true" enough to be saved, then he is true enough to use as a model for contemporary behavior—suggests as its own corollary a reciprocal relationship between the Christian present and the pagan past. The past bequeaths to the present models for virtuous action, and the present engages in acts of will and memory designed to manifest its appreciation and indebtedness. And God seems to approve, if his rescue of Trajan is any indication.

In the hagiographical tradition, as we have seen, God's approval comes on a case-by-case basis, and the admonition in the *Legenda aurea* that Gregory refrain from praying for any other damned soul shows that both God and the original author of the Trajan legend understood how the logic of this sort of request could lead to a bull market in intercessions. Virtuous pagan stories may begin with an exceptionalist premise—"*this one* merits reward for his virtues"—but acknowledging the exemplary status of one righteous heathen inevitably promotes the creation of a hypothetical *class* of exceptional individuals. If Trajan, why not someone else? Why not Cato, a question that occurred to Dante? Why not Vergil?[41]

That such logic occurred to and even appealed to Langland is clear from passus 12, where the subject of the virtuous heathen arises again. That Imaginatif should be the one to supervise the discussion is also quite logical; as the faculty in charge of marshalling absent images, past images, and even hypothetical images for rational scrutiny, Imaginatif is perfectly adapted to fill in the blank in the "if-Trajan-why-not-someone-else" question, which is exactly what he does.[42] Imaginatif certainly has a fideistic streak; he acknowledges that "Alle þe clerkes vnder crist ne kouþe þe skile assoile" why one thief as opposed to the other was saved at the Crucifixion: "Quare placuit? quia voluit" (B.12. 215–215a). At the same time, though, his reflection on the natural philosophy of the ancients leads exactly where we might expect it to, given the intercessory model sketched out in the previous passus.

> Swich tale telleþ Aristotle þe grete clerk. . .
> And wheiþer he be saaf or noȝt saaf, þe soþe woot no clergie,
> Ne of Sortes ne of Salomon no scripture kan telle.
> Ac god is so good, I hope þat siþþe he gaf hem wittes
> To wissen vs wyes þerwiþ þat wisshen to be saued—
> And þe bettre for hir bokes to bidden we ben holden—
> That god for his grace gyue hir soules reste,
> For letrede men were lewede yet, ne were loore of hir bokes.
>
> (B.12.270–76)

The "loore" of these ancients has already been successfully integrated into the learning of "vs wyes," and Imaginatif's sense of the debt thus owed— "For letrede men were lewede yet, ne were loore of hir bokes"—leads him to hope for their further integration, via the "reste" of salvation. He even suggests that our debt obliges us to pray for them, as Gregory did—or may have done—for Trajan. The difference between the way that Gregory "wilned to my soule / Sauacion for sooþnesse þat he seiȝ in my werkes" and Imaginatif's "hope. . .That god for his grace gyue hir soules reste" is really no difference at all, at least in terms of motive. Imaginatif's uncertainty derives solely from the lack of textual warrant in the cases he alleges ("no scripture kan telle"), not from any sense that Aristotle and company are less worthy of consideration. Indeed, what Imaginatif implicitly "imagines" here is an image of Aristotle et al. already saved—a concept that proves a little too much for Will, who falls back on what has become for him a familiar trope, a question about baptism.

Some forward progress has been made, however; this time Will's question is not about the sufficiency of baptism for salvation, as in the previous passus, but about its necessity.

> "Alle þise clerkes," quod I þo, "þat on crist leuen
> Seyen in hir Sermons þat neiþer Sarsens ne Iewes
> Ne no creature of cristes liknesse withouten cristendom worþ saued."
> "*Contra!*" quod Ymaginatif þoo and comsed to loure,
> And seide "*Saluabitur vix iustus in die Iudicij;*
> *Ergo saluabitur,*" quod he and seide na moore latyn.
> "Troianus was a trewe knyght and took neuere cristendom
> And he is saaf, seiþ þe book, and his soule in heuene.
> Ac þer is fullynge of Font and fullynge in blood shedyng
> And þoruȝ fyr is fullyng, and þat is ferme bileue:
> *Aduenit ignis diuinus non comburens set illuminans &c.*

Ac truþe þat trespased neuere ne trauersed ayeins his lawe,
But lyueþ as his lawe techeþ and leueþ þer be no bettre,
And if þer were he wolde amende, and in swich wille deieþ—
Ne wolde neuere trewe god but trewe truþe were allowed."

(B.12.277–90)

Imaginatif's speech finesses the issue of baptism's necessity by enlarging
Will's definition of "cristendom" to include multiple practices. The speech
has excited considerable commentary, largely focused on the kinds of baptism alluded to and the paean to "truþe" in lines 287–90, which is generally
taken to be a loose translation of the theological commonplace "facientibus
quod in se est deus non denegat gratiam," the well-worn phrase employed
in both scholastic texts and the work of the *moderni* to explain the salvation
of Old Testament patriarchs and other non-Christian *justi*.[43] "Truþe" is
of course one of the most resonant terms in the entire poem; Holy Church
avers three times in the first passus that "Whan all tresors arn tried treuþe is
þe beste" (B.1.85, 135, 207). It is also a word of considerable "polysemous
vitality" in both the poem and the period, as Richard Firth Green has
shown; "truth" was, according to Green, a Ricardian "keyword" in the sense
established by Raymond Williams, and for both Langland and his readers
the word would have operated in multiple and overlapping semantic
fields—legal, ethical, theological, and intellectual. Green cites the C-version
of Imaginatif's claim that "ne wolde neuere trewe god bote trewe treuthe
were alloued" (C.14.212) as "clear evidence that the poet was fully aware of
truth's semantic instability."[44]

Aware, and ready to exploit: the semantic density of "Ne wolde neuere
trewe god but trewe truþe were allowed" suggests a desire to take full
advantage of the word's overlapping senses and the permeability of the
boundaries between them.[45] In general, the salvation of the righteous heathen is where the desire to celebrate secular virtues and the need to preserve
doctrinal boundaries collide; that is to say, the whole concept derives from
the fraught collaboration of the ethical and theological imaginations (as the
criticism certainly makes clear). Imaginatif's lines turn out to be fully
fraught, achieving not just a sense of "truth's" instability but also a sense of
the word's plenitude of meaning and the opportunities for analogy it makes
available.

What, in the first place, does it mean to call God "trewe"? Certainly the
noun transfers back to the adjective something of the theological sense of
"truth," Truth as God, a sense that abides from its earlier uses in the poem
in the first passus (" 'The tour on þe toft,' quod she, 'truþe is þerInne,' " 1.12),
in the pardon scene ("TReuþe herde telle herof, and to Piers sente. . .,"
7.1), and elsewhere. But there is something of the legal sense of "troth" here

too: a "trewe" God can be trusted to fulfill his covenants, one of them apparently being his promise to reward "truth" with "a grete mede."[46] Finally, this willingness to faithfully and responsibly observe lawful covenants suggests the ethical sense of "truth"—that is, the kind of truth that characterizes Trajan and other true non-Christians and renders them worthy of attention in the first place. Recall Nancy Vickers's observation about the doubling of Trajan's function in the legend: "On one level, the emperor, acting like God, hears and answers the importunate request of the widow; and on another, God, acting like God, hears and answers the importunate request on Gregory's behalf." To call God "trewe" in *Piers Plowman*, then, is to suggest that he and Trajan share a concern for truth and the habit of practicing it—a suggestion confirmed by the use of "trewe" to modify "truþe" later in the line.

When it is used to refer to Trajan and those like him, "truþe" is obviously being used primarily in the ethical sense, to refer to the virtuous conglomeration of fidelity, integrity, justice and righteousness that makes Trajan a "trewe knyght." But the legal sense of the word is also lurking here, because of Trajan's reputation for "laweful domes"; when the lords are admonished to be mindful of Trajan's truth they are identified as lords "þat lawes kepe," those responsible for the application of truth as it is embodied in the law (another semantically loaded word, in fact). And in the context of Imaginatif's speech, with its emphasis on the "lawe" followed faithfully by the unbaptized true person, the "trewe truþe" of line 290 carries with it some of the weight of the theological sense, "truth" as a creed or set of beliefs.[47]

"Truth" and "law" are only two of the overdetermined, overloaded words in this passage. As Whatley demonstrates, the participle "allowed" can be glossed as both a secular word meaning "praised or commended" and as a technical theological term referring to the bestowal of salvific grace.[48] But perhaps the most suggestive word is "wolde," which connects not just the will of the true non-Christian (who *would* adopt a better, i.e., Christian belief if given the opportunity) with the will of God (who wills—or rather, would not will—the salvation of such an individual), but also links both of them with Gregory's will that Trajan be saved, and Imaginatif's hope for Aristotle, Socrates, and Solomon. Imaginatif's claim that "god is so good" is not very different from his assumption that God is "trewe," and in each passage what he implies is that the intercessory impulse Christians feel toward virtuous pagans—the will that desires their salvation—is analogous to God's will. Both are directed toward the same end, and though Imaginatif's uncertainty reminds us that the human will lacks the power to effect an exceptional salvation by itself, his hopefulness and the boldness of the analogy outlined here explains why he draws the conclusion he does from 1 Peter:" '*Saluabitur vix Iustus in die Iudicij, / Ergo saluabitur,*' quod he."

When the subject of extra-ecclesial salvation does arise again in the poem, the power of the divine will and the possibilities of human intercession are treated separately, in passages that can be individually powerful but that lack the daring and suggestiveness of Imaginatif's speech. As Watson has noted, the end of Christ's speech during the harrowing of hell seems to promise not only redemption for the baptized "but also the eleventh-hour rescue of all humankind,"[49] an act to be based on Christ's shared kinship with all people: "And my mercy shal be shewed to manye of my breþeren, / For blood may suffre blood boþe hungry and acale / Ac blood may noȝt see blood blede but hym rewe" (B.18.393–95).[50] The repetition of "blood" in these lines evokes the lexical density of Imaginatif's "trewe god" and "trewe truþe," linking the two passages; as Watson observes, "This crucial scene vindicates Will's earlier attempt to construct a theology of salvation acceptable to 'lewed' men by raising human need to the status of a universal principle, capable of coercing God into changing the course of salvation history. Sweeping aside the cautious accounts of salvation given by Imaginatif in favor of a 'reckless' response to the world's sin, Christ aligns himself with the most extreme voices in the poem, Will, Trajan, Patience, and Piers himself."[51] But Imaginatif's caution, once again, is based on the awareness that the human will can only hope for such exceptional rescues, not guarantee them; Imaginatif's power is limited merely to picturing the kind of scenario that Christ can actually cause to occur, through a recklessness that goes well beyond even the saintly Gregory's temerity in praying for a damned soul. Moreover, although the style and the sentiments are the same here as in passus 12, the absence of any human intercession or even participation registers the difference between the scenes. Will engages in debate and dialogue with Imaginatif, but at the Harrowing he is a silent observer, albeit an emotionally involved one.

In fact, the limits and proper orientation of human acts of intercession are described earlier than this, in Anima's long speech on charity, the church, and the conversion of the heathen in passus 15. Anima is not without a sense of history and offers an ecclesiological survey that pointedly if conventionally contrasts the purity and rectitude of the early church with its current depravity. Regarding non-Christians, however, Anima is quite present-oriented, departing from Imaginatif's concern with ancient philosophers and focusing instead on contemporary peoples like the Saracens and the Jews. Acknowledging the essential monotheism of these groups, Anima argues that since "þise Sarȝens, Scribes and Iewes / Han a lippe of oure bileue, þe lightlier me þynkeþ/ Thei sholde turne, whoso trauaile wolde to teche hem of þe Trinite" (B.15.501–03)—a position we will see again in *Mandeville's Travels*.[52] Conversion, then, is the new outlet for the intercessory impulse, and those who are most strongly obliged to "travail" to achieve it are the absentee bishops with sees *in partibus infidelium*, "Of Naʒareth, of Nynyue, of Neptalym and Damaske," "Of Bethleem and Babiloigne" (B.15.494, 510). Their models are Augustine, who brought the faith to the "hethynesse" of England and Wales, and the martyrs of the early church, who "In ynde, in alisaundre, in ermonye and spayne, / In dolful deþ deyeden for hir faith" (B.15.521–22). Thus the desire for the salvation of non-Christians, a powerful but unresolved concern in the poem, is here put to use in the interest of ecclesiastical reform, one nostalgic and universalizing impulse serving another.

Throughout his speech Anima stresses the need for non-Christians to be truly converted; the end of the passus even imagines them learning and repeating the elements of the creed. It is possible, of course, to read this emphasis on conversion as a retreat on Langland's part from the more liberal view of salvation articulated earlier, as "Langland's last—and much more conservative—word on the matter."[53] But it makes more sense to see Anima's remarks as marking one more point on a continuum that specifies different kinds of intercession for different circumstances: Gregory's act and Imaginatif's elaboration of a principle based upon it are concerned with the recuperation of the pagan past, while Anima's program of conversion is designed to solve the "problem" of contemporary non-Christians, and Christ's universalist impulse—the real last word—guarantees a merciful future. Such a schema has the Langlandian virtue of expanding the field of intercession to include both intellectual practice and the active life; some cases, like Trajan's, require acts of memory and prayer (and poetry), while others demand charitable and worldly deeds like preaching and teaching (and martyrdom). The distinctions are not purely objective ones, of course, based simply on who is dead and unable to convert, and who is still alive and thus still able to change (a contrast effectively undermined by *St. Erkenwald*, as we will shortly see). Langland himself shows different degrees of invest-ment, letting the salvation of Trajan stand alone while subordinating the conversion of the heathen to a larger program of ecclesiastical reform. Moreover, Trajan speaks, as do the philosophers through "loore of hir bokes," while the Saracens and Jews are spoken about, and are made the subject of some of the usual Christian slanders—Mahometseen a disappointed candidate for pope, for example. Still, it is possible to see Anima's speech about the insti-tution of the church as following the same route as Trajan's recommendations for individual reformation. Each of them extolls the salutary effects of poverty, though of course for Anima this means moving beyond individual ascesis to a more radical call for disendowment. Moreover, just as Trajan's peculiar pres-ence in the first inner vision implies that individual rejuvenation requires a turning outward, so Anima's program for ecclesiastical reform includes a turn toward those peoples outside the borders of Christendom.

". . .say me of þi soule. . ." ANIMA

Anima's speech on the conversion of the heathen marks a good place to move from *Piers Plowman* to *St. Erkenwald*, in part because the latter poem begins with a capsule account of Augustine's conversion of England, an accomplishment Anima also holds up as an example to the episcopal absentees of his own day. This is far from all they have in common; critics have long recognized the affinities between the alliterative *St. Erkenwald* and Langland's Trajan episode, generally judging the shorter poem to be the more conservative in its handling of the righteous heathen problem, a kind of theological backlash against what Whatley calls "the markedly secular character of the Gregory/Trajan story in the late medieval period."[54] But *St. Erkenwald* offers both less and more than just a reactionary revision of Langland's salvation theology; in the first place it is probably not so conservative as some have argued, and in the second we can find in *Erkenwald* a thorough recapitulation of all the forms and scenarios of intercession explored in *Piers*, including conversion, baptism, episcopal intervention, the trope of the talking pagan, memories of the harrowing, historical consciousness, and loaded terms like "law." In fact, *Erkenwald* consistently literalizes *Piers Plowman's* treatment of the righteous heathen theme, rendering as narrative what *Piers* typically offers as exposition.

The conservative reading of *St. Erkenwald* pointedly depends on the poem's narrative emphasis, for it is based on the fact that the author keeps his exceptionally righteous heathen alive long enough to partake of a romanticized though still orthodox baptism, and on the related fact that his saintly intercessor plays a more active role than even the Pope Gregory of the *Golden Legend*, to say nothing of *Piers Plowman*. The poem, Whatley claims, "presents the story in such a way as to magnify the role and prestige of the bishop and the visible sacramental church, which together turn out to be vital to the salvation of the heathen soul, regardless of his matchless justice."[55] In one sense this is indeed a conservative way to tell the story: compared to Imaginatif's plural approach to the issue or Christ's plans for his "breþeren," the recuperation of one pagan soul at a time is not particularly ambitious, and in *St. Erkenwald* the process requires an enormous investment in labor (the judge must be dug up, the tomb pried open) and episcopal attention. Baptism is essentially conservative as a technology, however we might regard it theologically: one at a time is the only way it works in medieval Christianity.

Beyond the question of mechanism, however, we can see how the poem is still deeply conditioned by the late-medieval "secular" understanding of the righteous heathen question, and its formal construction of that theme gives a good idea of just how the poet thought his public regarded virtuous pagans. In the formal structure and the narrative details of its virtuous pagan scene, *St. Erkenwald* points toward and ultimately delivers an answer to the righteous heathen question that is certifiably "secular," as indeed any answer must be that rewards earthly justice with an exceptional salvation. "*Ergo, saluabitur*," as Imaginatif might say. The probity of the pagan judge—indeed, his virtual entitlement to salvation—is quickly and firmly established. His justice is described in superlatives even beyond those Langland applies to Trajan.

> Bot for wothe ne wele ne wrathe ne drede
> Ne for maystrie ne for mede ne for no monnes aghe,
> I remewit neuer fro þe riȝt by reson myn awen
> For to dresse a wrange dome, no day of my lyue.
> Declynet neuer my consciens for couetise on erthe,
> In no gynful iugement no iapes to make
> Were a renke neuer so riche for reuerens sake.
> Ne for no monnes manas ne meschefe ne routhe
> Non gete me fro be heghe gate to glent out of ryȝt,
> Als ferforthe as my faithe confourmyd my hert.
>   (233–42)

Here is "truþ þat trespased never. . .ayeins his lawe," and the fulfillment of Trajan's command that lords "do truþe to the peple." Such is the *Erkenwald*-poet's acquaintance with the terms of the righteous heathen debate that he has personified the *facere-quod-in-se-est* ethic in the judge: not only did he adhere without deviation to the best law he knew, but as a judge he embodied that law. Its establishment and exacting, impartial enforcement are the very conditions of his being, the ontological underpinnings of "a lede of þe laghe" (200). Small wonder, then, that his body has been preserved by "þe riche kynge of reson þat riȝt euer alowes / And loues al þe lawes lely þat longen to trouthe" (267–68). These words—particularly the omnipresent "alowe"—might have been borrowed directly from Imaginatif's eulogy of Trajan.[56]

Bishop Erkenwald himself, in his questioning of the corpse, betrays exactly the kind of liberal understanding of the virtuous heathen question that such a testimonial might inspire. Surely, Erkenwald says, your soul must be in heaven?

> "ȝea bot sayes þou of þi saule," þen sayd þe bisshop,
> "Quere is ho stablid and stadde if þou so streȝt wroghtes?
> He þat rewardes vche a renke as he has riȝt seruyd
> Myȝt euel forgo the to gyfe of His grace summe brawnche,

For as He says in His sothe psalmyde writtes:
'Þe skilfulle and þe vnskathely skelton ay to me'.
Forþi say me of þi soule in sele quere ho wonnes
And of þe riche restorment þat raʒt hyr oure Lorde."
(273–80)

The Bishop is stunned to learn that the judge's soul is still in hell—and the judge is more than a little peeved himself: "Quat wan we wyt oure weledede þat wroghtyn ay riʒt, / Quen we are dampnyd dulfully into þe depe lake. . .?" (301–02). The querulous "we" betrays the kind of categorical thinking always implicit in the discourse of the righteous heathen, as we have seen; it suggests the possibility of a whole cohort left behind at the Harrowing, as the judge claims he was: "Quen þou herghedes helle-hole and hentes hom þeroute, / Þi loffynge out of limbo, þou laftes me þer" (291–92). Indeed, *St. Erkenwald* itself participates in this way of thinking, on the level of genre. While on the surface it ostensibly tells a unique tale, its clear affinities with the Trajan legend show that the poet was self-consciously working in a tradition analogous to, say, hagiography—many tales, but only one story, the story of the salvation of the heathen.

But the judge hasn't been saved yet, and of course his question is on its face susceptible to more than one possible answer. One reply would hew to Walter Hilton's position by critiquing the sense of entitlement implied in the judge's words. In the absence of grace, the argument would go, the correct answer is "nothing": virtuous acts may be meritorious in the secular sense but they simply confer no salvific benefits. The judge's question also recalls the one attributed to Dante by the Eagle in *Paradiso* 19: if a righteous man dies unbaptized on the banks of the Indus, where no word of Christ has ever been heard, "Where is this justice which condemns him? Where is his sin if he does not believe?" The answer of the Eagle could come from the Book of Job, as it responds not to the issue of equity but to questions of standing and jurisdiction: "Now who are you who would sit upon the seat to judge at a thousand miles away with the short sight that carries but a span?"[57] How is it that the pilgrim dares to ask such a question? But the implicit answer of the *Erkenwald*-poet is quite different and much more generous, recognizing in the judge's question a version of the one that prompts Christians to inquire after virtuous pagans in the first place. All the bishop can do in response is weep, and wish tearfully (and, coincidentally, in the exact words of the baptismal formula) for the judge's salvation. Thus, after carefully managing our expectations—and Erkenwald's—the poem graciously fulfills them. The righteous heathen is saved by the Bishop's almost inadvertent baptism, and his soul rises to heaven as his body crumbles to dust.

Thinking about literalization as the mode governing the *Erkenwald*-poet's response to Langland can bring a useful perspective to this debate over "liberal" or "conservative" accounts of virtuous pagans, and achieving some perspective on the issue may even have been what motivated the poet to narrate rather than simply recount an exceptional salvation. In *Piers*, of course, Trajan tells of his salvation, but Langland does not show it; it has already happened, and the fact that it has happened is what makes him the perfect respondent to Will's misguided claim about baptism. In *Erkenwald*, by contrast, the miracle hasn't happened yet, and the poet has to both stage-manage events and supply the necessary settings, characters, and props. Hence the intercessor, in this case Erkenwald, must be reintroduced into the tale, his lines restored, and he must have access to some mechanism for salvation, a speech or a gesture or a formula that will render the climax of the story accessible rather than mysterious. Seen from this more formal perspective, as having narratological as well as theological significance, the bishop's tears begin to seem overdetermined; thinking of them as a sort of prop or special effect frees us from the assumption that the tale is told wholly or largely for the sake of the baptismal moment, and lets us view the baptism as one of a number of devices—the tomb, the excavation project, the corpse—that lets the story work as it does.[58]

St. Erkenwald's strategy of literalization is not limited to the physical representation of an act of intercession; the action of the poem itself is a playing-out of some thematic concerns of *Piers*. The third vision's interest in the issue of soteriological equity are explored in *Erkenwald*, not only in the judge's sense of grievance ("What wan we wyt oure wele-dede. . .?") but in the person of the judge himself and his role—as a "lede of þe laghe" he is a literal embodiment of Trajan's own virtues, "trewe truþe" in the unde-composed flesh. And in a larger sense, *St. Erkenwald* is a literal response to Imaginatif's remark about how God wishes "trewe truþe" to be "allowed"; rather than offering a similarly subjunctive statement, in Imaginatif's mode, the *Erkenwald*-poet gives us the thing itself, turning hopeful speculation back into narrative. Moreover, the esteem Imaginatif expresses for Aristotle and other authors of indispensable "loore" is represented in the poem's very scenario: the church to which Erkenwald and his fellow citizens belong is literally founded on the body of a righteous heathen, one who himself embodies the laws and customs of his nation. A central irony of *St. Erkenwald* is the way it begins with the attempt of the "visible, sacramental church" to destroy the heretofore invisible foundations of the truth and justice that help to sustain its own culture. The poem's ultimate recuperation of those foundations through the judge's salvation is another suggestion that the *Erkenwald*-poet responds to *Piers Plowman* at the level of theme, not just incident; *St. Erkenwald* does not merely retell the story of Trajan's salvation in English guise, but rather explores the several contexts in which the virtuous pagan theme arises in Langland's poem.

Thus, the *Erkenwald*-poet may derive from *Piers Plowman* the idea that a sort of ethical genealogy connects the pagan past with the Christian present, but he gives the issue considerably greater and more explicit elaboration than does Langland. Much more than for *Piers*, the question of pagan virtue demands historical as well as moral investigation.[59] From the moment of the tomb's discovery its meaning is cast in terms of questions of history: who can this be? when could he have lived? and, most poignantly, why can't we remember his story?

> Þer was spedeles space to spyr vch on oþir
> Quat body hit myȝt be þat buried wos ther.
> How longe had he þer layne his lere so vnchaungit
> And al his wede vnwemmyd þus ylka weghe askyd:
> "Hit myȝt not be bot suche a mon in mynde stode longe;
> He has ben kynge of þis kythe as couthely hit semes.
> He lyes doluen þus depe hit is a derfe wonder
> Bot summe segge couthe say þat he hym sene hade."
> Bot þat ilke note wos noght for nourne none couthe,
> Noþir by title ne token ne by tale noþir
> Þat euer wos breuyt in burghe ne in no boke notyde,
> Þat euer mynnyd suche a mon, more ne lasse.
> (93–104)

The absence from the chronicles of any record seems most troublesome to the Londoners; when he explains the situation to Bishop Erkenwald, the Dean of St. Paul's laments that "we haue oure librarie laitid þes longe seuen dayes / Bot one cronicle of þis kynge con we neuer fynde" (155–56).[60] Indeed, it is this failure of memory that the Dean identifies as the true marvel of the body's discovery: "He has non layne here so longe, to loke hit by kynde, / To malte so out of memorie bot meruayle hit were" (157–58).

Criticism of the poem has tended to emphasize the distinction subsequently drawn by Erkenwald between faulty, incomplete human knowledge and divine intelligence, to which the people must now appeal for the solution to the mystery of the corpse's identity.[61] Erkenwald, however, interrogates not God but the corpse itself (albeit in God's name), and his questions recapitulate those of the crowd.

> Ansuare here to my sawe, councele no trouthe.
> Sithen we wot not qwo þou art witere vs þiselwen
> In worlde quat weghe þou was and quy þow þus ligges,
> How longe þou has layne here and quat laghe þou vsyt
> Queþer art þou ioyned to ioy oþir iuggid to pyne.
> (184–88)

When the revivified judge dates himself from the reign of "the bolde Breton Ser Belyn,"[62] it is by means of a complex (and probably scribally obscured) riddle that refers both forward from the founding by Brutus of New Troy—a favorite fourteenth-century name for London—and backward from the birth of Christ, "by Cristen acounte" (209). The judge inserts his story neatly into—and between—pagan and Christian history, just as he himself will shortly bridge the two cultures and be welcomed to the celestial banquet. When the historical status of the judge has been rendered intelligible—and only then—he can be worked into the Christian eschatology of the bishop and his fellow Londoners. Moreover, although the discovery of the tomb occurs in the course of the ongoing "historical activity and triumphant progress of the visible Christian church,"[63] the poem also acknowledges how that historical activity is continuous with the pagan activity that preceded it. St. Paul's was built, and is being rebuilt, on the site of the "temple...of Triapolitanes," "derrest of ydols praysid" in Saxon times—"in Hengyst dawes," before "Saynte Austyn into Sandewiche was sende fro þe pope..." (12).

Clearly, then, the "historical imagination" solicited by *St. Erkenwald* is itself a dynamic part of the poem, for it must reconcile the sometimes competing demands of church history, salvation history, and British history. As several critics have noted, one of the *Erkenwald*-poet's innovations is his creation of a homegrown righteous heathen, and the poem's thirty-two line prologue, in typical alliterative style, evokes a specifically British, specifically insular history.[64] Not only Sandwich and London, but Wales (9) and later Essex (108) are adduced to emphasize the Anglicization of the righteous heathen problem in this poem. And the attribution of the miraculous salvation to the intercession of Erkenwald also serves to anchor the poem's story firmly in the course of British history, for Erkenwald is a very British saint and no obscure one; his prominence is attested by historians and hagiographers from Bede through the fifteenth century. His shrine in St. Paul's was evidently sumptuous and the site of much miraculous healing, and proclamations enjoining the observance of his feast days (described as "of late neglected") issued in 1386 and 1393 show the saint to have been very much in the public consciousness of London in the late fourteenth century.[65] The *Erkenwald*-poet, then, takes the vernacularization of the righteous heathen well beyond Trajan's colloquial "Baw!" At the same time, though, the genealogy of Erkenwald's act can be traced back to Trajan's Gregory; Gregory sent Augustine to convert the British, as Anima notes, and Erkenwald is his direct spiritual heir: "of þis Augustynes art is Erkenwolde bischop" (33).

In treating the Erkenwald-poet as Langland's heir we can almost see him working his way backwards through *Piers Plowman*, beginning with Augustine and the topic of conversion (one of Langland's themes in passus 15), exploring the issues of equity and merit relevant to salvation (as Imaginatif does in passus 12), and concluding with the exceptional salvation itself (Langland's starting point, with Trajan). But the poem's reading of *Piers* is not limited to that poem's pagans, for it also alludes to and literalizes the activity of Langland's Meed episode, specifically the intervention of Conscience and Reason in passus 3 and 4. In decrying Meed's corruption of justice and the law, both of these figures are ultimately reduced to a kind of apocalyptic riddling that looks forward to an utterly reformed polity in which, as Conscience puts it, ". . .Reson shal regne and Reaumes gouerne, / Shal na moore Mede be maister on erþe, / Ac loue and lowenesse and leaute togideres; / Thise shul ben Maistres on moolde trewe men to saue. / And whoso trespaseþ to truþe or takeþ ayein his wille, / Leaute shal don him lawe and no lif ellis" (B.3.285, 290–94). Indeed, Conscience's prophecy even involves the conversion of the heathen:

> And er þis fortune falle men fynde shul þe worste
> By six sonnes and a ship and half a shef of Arwes;
> And þe myddel of a Moone shal make þe Iewes torne,
> And Sarȝynes for þat siȝte shul synge *Gloria in excelsis &c.*,
> For Makometh and Mede myshappe shul þat tyme. . .
> (B.3.325–29)

Reason engages in the same reformist fantasizing in the next passus, imagining a perfectly comprehensive and equitable justice:

> I seye it by myself, and it so were
> That I were kyng with coroune to kepen a Reaume,
> Sholde neuere wrong in þis world þat I wite myȝte
> Ben vnpunisshed at my power for peril or my soule,
> Ne gete my grace þoruȝ giftes, so god me helpe!
> Ne for no Mede haue mercy but mekenesse it made,
> For *Nullum malum* þe man mette wiþ *inpunitum*
> And bad *Nullum bonum* be *irremuneratum*.
> Late þi Confessour, sire kyng, construe it þee on englissh. . .
> (B.4.137–45)

Construe it in English is just what the author of *St. Erkenwald* does, offering in the pagan judge a literal embodiment of the perfect justice described here, in language that specifically echoes the Meed episode of *Piers*:

> Þe folk was felonse and fals and frowarde to reule,
> I hent harmes ful ofte to holde hom to riȝt.
> Bot for wothe ne wele ne wrathe ne drede

Ne for maystrie ne for mede ne for no monnes aghe,
I remewit neuer fro þe riȝt by reson myn awen
For to dresse a wrange dome, no day of my lyue.
Declynet neuer my consciens for couetise on erthe,
In no gynful iugement no iapes to make
Were a renke neuer so riche for reuerens sake.
Ne for no monnes manas ne meschefe ne routhe
Non gete me fro þe heghe gate to glent out of ryȝt,
Als ferforthe as my faithe confourmyd my hert.
Þaghe had been my fader bone, I bede hym no wranges,
Ne fals fauour to my fader, þaghe felle hym be hongyt.
                                                    (231–44)

In this passage Langland's allegorical characters return as common nouns, "mede" and "reson" and "consciens," and in the judge the *Erkenwald*-poet offers a model of juridical rectitude that, in an environment much like that described in *Piers* ("þe folk was felonse and fals. . ."), resists the corrupting influence of meed and leaves no malefactor unpunished, regardless of rank or relation. The judge's first reward—the citizens of New Troy "coronyd me þe kidde kynge of kene iustices" (254)—literally if posthumously enacts what Langland's Reason can only imagine conditionally and prophetically ("That I were kyng with coroune to kepen Reaumes"). And his second reward, his miraculous salvation, carries out on a modest individual scale the kind of apocalyptic transformation Conscience calls for as a cultural imperative, in which the corrupted secular discourse of the law will be reformed spiritually when "Reson shal regne" and "oon Cristene kyng kepen vs echone" (B.3.289). For Conscience the time of this transformation can only be a riddle; *St. Erkenwald*, by contrast, begins with a riddle—who is this that we have discovered?—and proceeds to solve it, concluding not with the breakdown of court satire as in *Piers* but with a glimpse of the heavenly court "þer soupen alle trewe."[66]

What *St. Erkenwald* seems ultimately to take from *Piers Plowman*, and largely from Imaginatif, is a sense that the work of intervention is never done. Salvation for the just non-Christian, the only equitable and in fact only imaginable solution to the so-called virtuous pagan problem, relies on constant acts of intercession and reciprocal honor done in the present, in the name of those in the past. Their salvation, like Trajan's, is contingent upon "our" willingness, like Gregory's, to honor them as worthy models. *Erkenwald* moves beyond *Piers* in its implicit recognition that ongoing acts of intercession can be textual rather than miraculous, and that rewriting the Trajan story is a way of reenacting it, and encouraging yet more reenactments. Vickers, in her study of Dante's use of the Trajan story in the *Purgatorio*, has noted that "within the drama of intercession there may well be a third term: intercessor, intercessee, and that which moves one to intercede—the work of art, the text" (80). *St. Erkenwald* suggests that there is a fundamental analogy between the text that motivates intercession and that which later documents it, and that in the course of time and in the process of historical and cultural *translatio* that second text—a chronicle like *St. Erkenwald*—takes on the power of the first and becomes itself the third term, mediating between a new audience and its desired (and deceased) model of virtue. Ultimately, the eschatological recuperation of virtuous pagans—their rescue from hell to heaven—is not only perfectly analogous but perfectly identical to the poetic and historical recuperation—their inclusion in texts—that brings them from obscurity and darkness into the light of contemporary public consciousness.

Here again *St. Erkenwald* is self-consciously emblematic, because the identity and meaning of the judge is plainly figured as a textual problem, not only because of his inexplicable absence from the chronicles and the "martilage" of St. Paul's, but because of the letters that no one can read written on his magnificent tomb. The sepulchre, like a page from an illuminated manuscript, has its

> bordure enbelecit wyt bryȝt golde lettres,
> Bot roynyshe were þe resones þat þer on row stoden.
> Fulle verray were þe vigures þer auisyde hom mony,
> Bot alle muset hit to mouthe and quat hit mene shulde:
> Mony clerkes in þat clos wyt crownes ful brode
> Þer besiet hom a-boute noȝt to brynge hom in wordes.
>
> (51–56)

The "roynyshe" figures are never deciphered or translated in the poem, I would argue, because they are a figure for the poem itself, which supplies the deficiencies of chronicles and corrects the failures of memory that render the letters unreadable.[67] In fact, the poem acknowledges its own function just prior to the discovery of the "ferly": "And as þai makkyde and mynyde," the poet says of the workers, "a meruayle þai founden / As ȝet in crafty cronecles is kydde þe memorie. . ." (43–44). But the chronicles of London and Saint Paul's lack any records of the event: *St. Erkenwald* is the "crafty cronecle" that describes the marvellous discovery and the even more marvellous events which ensue.[68]

All those events are notably public acts in *St. Erkenwald*, witnessed by the bishop, the mayor, the Dean of St. Paul's, lords and barons and "mony hundrid hende men." In contrast to *Piers*, which stages its virtuous pagan episode as an acutely private, interior moment, as the inmost part of an inner vision, *Erkenwald* brings it all into the open, representing within the poem an audience that is clearly analogous to its fourteenth-century English audience. The standard reading of the poem argues quite rightly that both audiences are there to be edified by the exchange between bishop and judge, but there is more going on than just instruction; within the poem (and presumably without) the audience is also being moved, moved to tears, and enlisted into a regime of sympathy for those who lived with Pelagian expectations but seem at first to have received only Augustinian compensation.

   In Langland's poem that sympathy is most fully felt by Imaginatif, as we have seen, and in his redaction of the Trajan legend the poet of *St. Erkenwald* uses his own *vis imaginativa* to combine two of Imaginatif's profound concerns, the fate of the righteous heathen and the moral status of the "makynges" with which Will keeps meddling. If *St. Erkenwald* proves anything, it is that even in the wake of *Piers Plowman* there were not nearly "bokes ynowe" about pagan virtue.[69]

CHAPTER 2

MANDEVILLE'S "GRET MERUAYLLE"

To paraphrase Sir John himself, there are two ways to begin a critical account of *Mandeville's Travels*: some start by remarking upon the book's extraordinary medieval popularity, as evidenced by the hundreds of surviving manuscripts and incunabula and its early translation into multiple European vernaculars, while others prefer to discuss the book's critical reputation, paying particular attention to the sneering dismissals of its nineteenth- and early twentieth-century editors, whose work on the book's sources led them to brand the *Travels* a hoax or a plagiarism.[1] But both of these approaches, like different routes to Jerusalem, lead essentially to the same place, a loose but consistent critical consensus that may not be able to agree on exactly what the book *is* (a "romance of travel?" a satire? a geography? pre-modern prose fiction?[2]) but generally agrees on the attitude it expresses: a genial, ironic, worldly wisdom that exhibits a surprising and thoughtful interest in, and a hedged approval of, the practices, beliefs, and virtues of non-Christian, non-Western peoples.[3] The texts of the *Travels* provide ample evidence for this view—so much, in fact, that *Mandeville's Travels* has more often been described than analyzed by its scholarly readers over the last half-century. The text, through its undogmatic voice, its paratactic style, and its ungovernable form, often seems to cast a spell over those who would study it such that they adopt its conclusions as their own: that the world, though fallen, is nevertheless blessed through its widespread acknowledgment of one god; that pious devotion and fidelity to one's beliefs outrank mere doctrinal or theological correctness; that borders are not impassable, boundaries of all sorts are not absolute, and that all of us, made wise by travel and experience of the world (even if only vicariously accomplished), can get along.

Recent and important modifications of this assessment have called attention to the anti-Jewish sentiments of the text, the dark side of its apparently tolerant embrace of diversity, and while such sentiments are certainly not untypical of medieval writing it is somewhat jarring to discover them in a text that has traditionally been seen to correspond so well to modern liberal latitudinarianism.[4] This rediscovery of Sir John Mandeville's anti-Semitism is potentially a very salutary development for study of the *Travels*; attention to the text's premises (i.e., the irremediable exclusion of the Jews from Mandeville's unified vision of the world) rather than its conclusions (i.e., that the world can be seen as somehow unified) can provide some critical leverage with which we might dislodge, if not entirely discard, some Mandevillean pieties. In this chapter I argue that the idea of pagan virtue in *Mandeville's Travels* is also one of its premises, an organizing principle rather than an enlightened conclusion. The *Travels* achieves its effects and promotes its apparently humane view of the world through a nearly perfect marriage between the conceit of travel and the principle of pagan virtue, which mutually reinforce one another throughout the text.

Travel is the conceit that lets "Sir John Mandeville" offer his representation of the world as something other than a scholarly summation, in the same way that Chaucer's conceit of a Canterbury pilgrimage allows him to organize his collection of tales according to "what really happened" as opposed to some obvious authorial scheme. Though we now recognize "Sir John" as a fictional persona created to give continuity to a compilation derived from almost three dozen other texts, such is the power of the travel conceit that medieval readers took him to be a real English traveler and the author of an authentic account of his own adventures. Moreover, though the narrator's own Englishness is a (perhaps significant) fiction, the text's appeal to the English imagination was very real; there are more English versions of the *Travels* than there are French, Latin, or Dutch/German recensions (though there are more extant manuscripts in those languages).[5] Sir John even occasionally appears in other people's books: a chronicler in the abbey of Meaux in Yorkshire, writing probably in the last decade of the fourteenth century, includes in his entry for 1356 not only the battle of Poitiers but also the composition of the *Travels* by "Johannes de Mandavilla, miles Anglicus, in villa Sancti Albani oriundus," a book that the chronicler claims (uniquely, I believe) was dedicated to Edward III.[6]

Borrowing from the prologue of the *Travels*, the Yorkshire chronicler notes how Sir John traveled through lands inhabited by "diversae gentes diversorum rituum et formarum," what the Cotton texts refers to as "many dyuerse folk and of dyuerse maneres and lawes and of dyuerse schappes of men" (prol. 3).[7] The summary emphasis on diverse places and the diverse people that occupy them anticipates the rhetoric of many modern critics, but it also suggests how for Mandeville and his contemporary readers the conceit of travel can relocate the topic of pagan virtue from the theological realm to the empirical world; when Sir John praises the righteousness of a

Saracen or a virtuous Brahman he claims to do so out of his direct experience of that righteousness, rather than according to some abstract doctrinal speculation or patristic commonplace. *Mandeville's Travels* thus insists (in its gentle, humane, ironic way, of course) that its readers think of righteous heathens not as eschatological gate-crashers but as contemporaries, as a topic of vernacular geography as well as vernacular theology. At the same time, the text's recurrent foregrounding of devout, rational, praiseworthy non-Christian folk, whether Cathayan or cannibal, serves to hold together, albeit often loosely, a journey that exhibits from the start strong centrifugal tendencies, an adventure driven by the loosest geographical goals (the *Travels* consistently moves from west to east, but certainly not steadily or logically so) and the obscurest of motives (we learn right away why Sir John wrote of his travels—to satisfy the curiosity of his western Christian readers—but never specifically why he undertook his putative expedition in the first place).

In what follows I first focus on three different virtuous pagan scenes in the *Travels*, Mandeville's accounts of—or rather, encounters with—the Saracens and their Sultan, the Cathayans and their Great Chan, and the "Bragmans" of India. Each of these sections—they appear in chapters 15, 23–25, and 32 of the Cotton text—echoes the others, as each makes similar observations about the moral intelligibility and earthly rewards of virtuous pagan behavior. Moreover, each passage exhibits the evolving, reciprocal relationship between the conceit of travel and the principle of pagan virtue, for in each case it is the fact of travel itself that legitimizes both Sir John's claims and the pagan self-representations on which they are based. That is, while each episode offers some domesticating rhetoric, some prophecy or shadowy Christian trait that tends to render its heathen subject less threatening or more familiar to an orthodox Western reader, each also contains some moment of empirical insight that tends to reassert more disturbingly heterodox conclusions. In the last section, I argue that the final chapter of the Cotton text of the *Travels* represents a particularly English response to the theme of pagan virtue as it appears throughout the text; the so-called papal interpolation in that version (and other English redactions) turns the last chapter of the *Travels* into a perfect summary of the whole, one that seeks to resolve (but ends up recapitulating) the spiritual and formal anxieties that are intertwined throughout the text.

Egypt: East Is West, and West Is East

The *Travels'* account of the Saracens is among the most positive in medieval Western writing; Mandeville's description of "Saracen" beliefs, derived largely from William of Tripoli's *De statu Saracenorum* (1273), repeats few of the conventional medieval slanders against Islam and stresses, as one might expect, the doctrinal proximity of the Muslim and Christian faiths.[8] And although his history of the Egyptian sultanate in Chapter 6 is a litany of parricide, fratricide, and assassination, the climax of his account—Sir John's interview with the Sultan in Chapter 15—presents a very different view of the present, that is, of the Saracens of Sir John's own acquaintance.

Sir John prepares for this extraordinary private colloquy by previously establishing for us his close relations with the Sultan: "I duelled with him as soudyour in his werres a gret while ayen the Bedoynes," he claims in Chapter 6, adding that "he wolde haue maryed me full highly to a gret princes doughter yif I wolde han forsaken my lawe and my byleue, but I thanke God I had no wille to don it for no thing that he behighte me" (6.24). Later he asserts that he was able to travel freely in the Sultan's lands because "I hadde lettres of the Soudan with his grete seel, and comounly other men had but his signet; in the which lettres he commanded of his specyalle grace to alle his subgettes to lete me seen alle the places and to enforme me pleynly alle the mysteries of euery place and to condyte me fro cytee to cytee. . ."(11.60). Having established himself as a favorite and intimate of the Sultan, whose special favor is here used to underwrite the text's own ethnographic project, it is thus perfectly reasonable that Sir John would pause to illustrate his observations on the Saracen faith with a revealing personal anecdote (derived from Caesarius of Heisterbach's *Dialogus Miraculorum* [1223][9]). That the rebuke which ensues closely resembles the dialogue between one Brother William of Utrecht and a Saracen emir from Caesarius's early thirteenth-century collection of *exempla* does not diminish the originality of the author of the *Travels* in inserting it into his already sympathetic account of Saracen beliefs, and the flourish with which he ends the episode renders it one of the most arresting passages in the book.[10]

The passage begins after Sir John has concluded that the Saracens "han many gode articles of oure feyth, alle be it that thei haue no parfite lawe and feyth as Cristene men han" (15.100). The Saracens are monotheists, believing in the Creation and Doomsday, and they venerate both the Virgin Mary and Jesus, whom they acknowledge as the Word of God. In fact, they might be easily converted due to this proximity to the Christian law, although they are mired, he complains, in a literal rather than spiritual understanding of the Scriptures. And here he digresses, that the Christians might be cursed in person:

> And therfore I schalle telle you what the Soudan tolde me vpon a day in his chambre. He leet voyden out of his chambre alle maner of men, lordes and othere, for he wolde speke with me in conseille. And there he asked me how

the Cristene men gouerned hem in oure contree, and I seyde him, "Right wel, thonked be God." (15.100)

The convention that underwrites Mandeville's hapless reply renders it utterly disingenuous. When confronted with a similar question in the *Dialogus Miraculorum*, William of Utrecht, "unwilling to say what the truth was," played it cagily: "Well enough," he answered, "*Satis bene*."[11] Here, as there, it precipitates a diatribe excoriating everything from Christian gluttony, irreverence, belligerence, deceit, avarice and lechery to their prideful slavery to fashion: "thei knowen not how to ben clothed, now long, now schort, now streyt, now large, now swerded, now daggered, in alle maner gyses" (15.100). And the Sultan firmly attaches his rebuke to the prevailing theme of sin and dominion already established by Sir John in the prologue, where he himself had described the sinfulness and depravity of Western Christian lords as responsible for the loss of the Holy Land.[12] The Sultan agrees:

> And thus for here synnes han thei lost alle this lond that wee holden. For for hire synnes here God hath taken hem into oure hondes, noght only be strengthe of oureself but for here synnes. For wee knowen wel in verry soth that whan yee seruen God, God wil helpe you, and whan He is with you, no man may ben ayenst you. (15.101)

Two observations follow from this remark of the Sultan's. The first is that however much the Sultan's words may look like simple ventriloquism on the part of the author, the use of a satirical stand-in to deliver some self-criticism, they are really something more. For Mandeville's straightforward and conventional political theory—that lords must live well to rule well—applies to the Sultan as well as to Western Christian lords. And while the latter have failed the test due to their "pryde, couetyse and envye" (prol. 2), the Sultan has passed it with flying pennants: he and his people are "gode and feythfulle" in obeying the commandments of "the holy book Alkaron" (15.102), and he is thus a ruler of great power and magnificence. What is offered as a proposition in the prologue, an equation between moral probity and political success, is here proven, cross-culturally, as a law. For Sir John the traveller has seen the evidence in the Sultan's realm: his castle, the five kingdoms under his rule, the thousands of men-at-arms he commands, his loyal "amyralles," the elegance and richness of his court, and the reverence paid to him there (all of which merely previews the magnificence surrounding the Great Chan) all depict a powerful and secure lord whom Mandeville seems proud to have served. The description peaks with a passage about the Sultan's magnanimity: no stranger who makes a request of the Sultan will be denied its fulfillment, provided that it does not contradict Saracen law. Indeed, all the princes under the Sultan's rule adopt this policy in imitation of him, saying that "no man schalle come before no prynce but. . . schalle be more gladdere in departynge from his presence thanne he was at the comynge before hym" (6.28).[13] In essence, the Sultan is everything that the Christian lords of the prologue are not, and if for us the account of his kingdom is a sort of projection of a fallen Western ideal onto a Saracen "screen," its effect within the text is to invest the Sultan with the moral authority to make his judgments against his Christian peers. It proves that the relationship between political and moral authority is not only rhetorical, but reciprocal, and its evidence for that proof is the evidence of Sir John's own eyes.[14]

At the same time, the Saracens' moral authority is subject to several conventional strategies of containment, and this leads to the second observation about the Sultan's indictment. He may be right, and his words may be cause for amendment, but they are not cause for alarm, because Islam is not a threat to the West. Indeed, the very proximity of the tenets of Islam to Christian doctrine undermines the former's claims to integrity and self-sufficiency, as if by some process of supernatural selection it will simply evolve into an orthodox Christianity. The author of the *Travels* inherits this tactic from William of Tripoli, who devotes more than half of the *De Statu Saracenorum* to cataloguing the contiguities of Christianity and Islam, concluding hopefully from their proximity that "through the simple word of God, without philosophical arguments or military warfare, like simple sheep [the Saracens] will seek the baptism of Christ and pass into the sheepfold of God."[15]

Further circumscribing the moral threat of Islam is the strategic deployment of prophecy throughout the *Travels*, also a technique it shares with the *De Statu Saracenorum*.[16] Evidently even the Saracens believe the Christians destined to retake the "Lond of Promyssioun," when (not if) they amend. The Sultan admits to Sir John "that knowe we wel be oure prophecyes that Cristene men schulle wynnen ayen this lond out of oure hondes whan thei seruen God more deuoutly" (15.101). Indeed, according to Mandeville Islam itself shall vanish according to prophecy: "And also thei seyn that thei knowen wel be prophecyes that the lawe of Machomete schalle faylen as the lawe of the Iewes dide, and that the lawe of Cristene peple schalle laste to the Day of Doom" (15.98–99). The recourse to prophecy allows the author of the *Travels* to assert the moral successes of the virtuous heathen while yet containing them within a larger Christian context, and ultimately an orthodox one. Indeed, speculation about the inevitable if historically indeterminate (in fact, infinitely deferrable) demise of Islam not only diminishes the urgency of the call for a crusade or an elaborate program of proselytizing, it also reduces the duty of Christians interested in the intractable problem of the Saracens to the mere maintenance of their faith, which is bound to triumph by its very nature. All a good Christian really has to do is not become a Saracen. Thus, Sir John's own orthodoxy functions as a sort of boundary for the exposition of Saracen virtues: although he is a self-confessed admirer, he is firm enough in his own faith to dismiss the thought of converting to their law, despite the offer of marriage to "a gret princes doghter."[17] Although the *Travels* begins with the rhetoric of crusade and asserts that "euery gode Cristene man that is of powere and hath whereof scholde peynen him with alle his strengthe for to conquere oure right heritage [i.e., the Holy Land] and chacen out alle the mysbeleeuynge men" (prol. 2), the text ultimately evolves to a position that is equal parts optimism and quietism. As Geraldine Heng observes, "Successful travel romance answers well to one crucial question in cultural colonization: How might it be possible to act at a distance on far-flung objects, places, peoples, and events? The answer—by bringing those distant places and peoples back home in such a way that the very process of retrieval itself would work to identify 'home' as a center—is the particular specialty, and province, of travel romance," which seeks to "insert its audience into an overarching system in which domination at a distance can seamlessly and seductively occur."[18]

The author of the *Travels* thus appears to make the Sultan's critique both more than the allegorized fulmination of a conventional moralist by investing him with the ethical authority to speak freely and legitimately, and less than a true threat to the moral integrity of Christianity by creating an overarching structure of prophecy to provide an orthodox, if unusually positive, perspective. But Sir John's amazing conversation is not over yet.

To return to the Sultan's chamber: when Sir John asks the Saracen chief where he acquired the information upon which his accusations are based, the latter replies that "he knew alle the state of alle contres of Cristene kynges and princes and the state of the comounes also be his messangeres that he sente to alle londes, in manere as thei weren marchauntes of precyous stones, of clothes of gold, and of othere thinges, for to knowen the manere of euery contree amonges Cristene men"(XV.101). Chaucer's Man of Law calls merchants "fadres of tidynges" in his prologue, and the author of *Mandeville's Travels* shows us why. They're spies:

> And than he leet clepe in alle the lordes that he made voyden first out of his chambre, and there he schewed me iiii. that weren grete lordes in the contree, that tolden me of my contree and of manye other Cristene contrees als wel as thei had ben of the same contree, and thei spak Frensch right wel and the Sowdan also; whereof I had gret meruaylle. (15.101)

Sir John's "meruaylle" is well-founded; this is a strikingly unconventional moment in the *Travels*, for a number of reasons. In the first place there are the adaptations that the author of the *Travels* has made to this episode as it comes down to him from Caesarius of Heisterbach. In the *Dialogus Miraculorum*, the Saracen emir who is William of Utrecht's companion claims to have learned French in the household of the King of Jerusalem, where he had been sent as a boy in a sort of aristocratic cultural exchange program; his indictment of Christian behavior is limited to the inhabitants of the Kingdom of Jerusalem, although Brother William's "satis bene" clearly implies that the rebuke is meant for Christendom at large.[19] But in the *Travels* what had been open cultural exchange becomes covert infiltration, disguised as trade and issuing in direct knowledge of Christian depravity. The Sultan's critique of Western mores is not implied, but bluntly stated, and based on immediate and empirical evidence—the testimony of his spies. Thus, over and above his carefully and logically elaborated moral authority, the Sultan has a practical and empirical right to speak about the vicious habits of the Christian West. The superstructure of prophecy and doctrinal proximity created to limit both the integrity and the evaluative power of Islam cannot restrict evaluations based on direct observation—not in a text the very conceit of which is the collection and assessment of information through direct ethnographic study, through the very adventures and travels of the eponymous Sir John Mandeville. If we accept the premise of *Mandeville's Travels*, we must also accept this fundamental analogy between one kind of discovery and another. It is perhaps no coincidence that one Latin word for spy—"explorator"—has come to mean something quite different for us.

And here arises a second problem. For not only is Sir John like the Saracen merchant-spies in his fact-gathering activities, but the occupation that enables him to travel and learn—mercenary knight—is analogous to theirs as well. Later in the century, Philippe de Mézières—perhaps drawing on Mandeville—would claim that "the spies through whom one might best learn of the state of his enemies are Lombards and other foreign merchants, who because of their trade have occasion in person or through their agents to send from one part to another; and especially those merchants who by means of their trade in precious stones gain access and private friendship with kings and princes."[20] Like De Mezieres's Lombards, Sir John's private friendship with the Sultan has earned him the privilege of inquiring into "alle the places" and of exploring "pleynly alle the mysteries of euery place" in the Sultan's domain. And it is through these investigations and this participation in the daily business of the Arab world, speaking we must presume adequate Arabic (for Sir John claims to have read the Koran, and the Sultan evidently does not break into French until his espionage network has been revealed[21]), that Sir John's own compatriots learn about the Arabic world, and beyond. Who else knows French "right wel" in *Mandeville's*

Travels? Why, Sir John himself, of course: "And yee schulle vndirstonde that I haue put this boke out of Latyn into Frensch and translated it ayen out of Frensch into Englyssch, that euery man of my nacoun may vndirstonde it" (prol. 3–4).

Rather than precious stones, however, Sir John exchanges his military services for information and the chance to study the world through which he moves. That the narrator of *Mandeville's Travels* is a knight is the first thing we learn about him: "I Iohn Maundevylle knyght, alle be it I be not worthi, that was born in Englond in the town of Seynt Albones, and passed the see in the yeer of oure lord Ihesu Crist m.ccc. and xxii. in the day of Seynt Michelle, *etc., etc.*" (prol. 3). We soon learn that that his knighthood is more existential than it is feudal, that it comes with an allegiance that can be transferred or sold. It crosses national boundaries not because of some inherent quality of transcendence, but because it stands for a set of valuable and sought-after skills—because it can be commodified and rendered as negotiable as a gem and as marketable as cloth of gold.[22]

Hence Sir John's service in the Sultan's wars against the "Bedoynes" in Chapter 6 (like his later fifteen-month stint in the Great Chan's army "ayenst the kyng of Mancy" in Chapter 23) demonstrates that the struggles for succession and the wars of conquest whose history he relates throughout the *Travels* extend into the present as well, a present in which Sir John is meant to be thoroughly implicated, not as a neutral observer or an itinerant romance knight, but as an historical actor. In other words, the exploration of pagan virtue through the conceit of travel reveals that it is not interdenominational devotion to the worship of the one God but international complicity in the realm of trade, war, and politics that connects the Christian West to the Saracen East. That this is a matter of considerable "meruaylle" is indicated by Sir John's reaction. Rather than expressing alarm at the extent of Saracen espionage, he expresses dismay that said spying should underwrite the Sultan's denunciation of Christian mores. "Allas," he laments, "that it is a gret sclaundre to oure feith and to oure lawe, whan folk that ben withouten lawe schulle repreuen vs and vndernemen vs of oure synnes" (15.101). This remark is in one way entirely conventional; although it implicitly accepts both the justice of the Sultan's charge and the practical success of the Saracens' covert operations, it tries at the same time to reassert the doctrinal hierarchy that places the Christian "lawe" over the Saracen one. We should be converting the Saracens by our good example, Mandeville argues, because our faith is superior. Yet, even this attempt to restore conventional Christian values is compromised by Sir John's own knowledge of Muslim behavior: "And treuly thei sey soth. For the Sarazines ben gode and feythfulle, for thei kepen entierly the commandement of the holy book Alkaron that God sente hem be His messanger Machomet. . ." (15.102).[23]

Thus cornered, confronted by the logical conclusions of his own explorations, Sir John changes the subject. He gives a sketch of the life of "Machomet," ending with his second retelling of the legendary tale of the Muslim prohibition against wine.[24] While the prophet lies in a drunken sleep, his companions use his sword to slay a Christian hermit friend of his, whose preaching had evidently tended to keep them up all night. When Machomet awakes and discovers the body, his men show him his own bloody sword and swear that he committed the act in a drunken state. Machomet curses wine and all those who drink it; therefore, Sir John explains, devout Saracens never drink wine. At least, not publicly: "summe drynken it preuyly, for yif thei dronken it openly thei scholde ben repreued" (XV.103).

These anecdotes about Saracen perfidy and hypocrisy seem at first to be a recuperative gesture designed to offset the Sultan's account of Christian moral failure; if the Christians can't be defended, at least the Saracens can be diminished. But what happens is the opposite, because what Sir John next reveals is that "it befalleth sumtyme that Cristene men becomen Sarazines outher for pouertee or for sympleness or elles for here owne wykkednesse" (XV.103). Thus, a chapter that begins with the high hope of easy conversion of the Saracens, because of the contiguity of Christian and Saracen belief, threatens to end with an image of exactly the opposite: Christians converting to Islam. And once again Sir John changes the subject. The last paragraph of the chapter is also the last paragraph of the first half of the book, the portion devoted to Sir John's tour of the Holy Land (that is, the portion based on William of Boldensele's *Itinerarius*, though the bulk of the Saracen material comes from William of Tripoli). In it Sir John describes the Saracen alphabet, which he reproduces, just as he had earlier reproduced the Greek (Ch. 3), Egyptian (Ch. 7), and Hebrew (Ch. 12) scripts. In a departure from the previous passages, however, Sir John here offers a bit of alphabetic analysis. He observes that

> iiii. lettres thei haue more than othere for dyuersitee of hire langage and speche, for als moche as thei speken in here throtes. And wee in Englond haue in oure langage and speche ii. lettres mo than thei haue in hire abc, and that is þ and ȝ, the whiche ben clept thorn and yogh. (XV.104)

M.C. Seymour notes quite rightly that the citation of thorn and yogh are "part of [Mandeville's] attempt to create a credibly English character."[25] But the placement of this apparent grace note here, at the end of a chapter that has dealt at length with the differences between Christian and Saracen, gives it additional weight. It suggests that the mere fact of difference ought not always to imply the necessity or immanence of moral evaluation. Sometimes it's just alphabetical: the more guttural Saracen language quite practically and logically requires different symbols. By citing a pair of unique English symbols without supplying the particular rationale for their use (i.e., to represent sounds not present in Latin and not adequately captured by the Roman alphabet), Mandeville avoids implying the existence of any hierarchy of tongues, and suggests, in keeping with his overall ethnographic approach, that the differences between languages are natural, almost trivial adaptations to local circumstances.[26]

Thus the question of what difference means—doctrinal difference, moral difference, linguistic difference—is left deeply unsettled at the end of this chapter, as is the relationship between one kind of difference and another. Christianity may be superior to Islam, yet Christians are the ones converting to the Saracen faith. The Saracens live devout and moral lives and are clearly more righteous than their Western counterparts, yet even the Saracens' own prophecies foretell their downfall at the hands of a morally rejuvenated Christendom. Finally, Sir John offers up a fact of difference—the diversity of languages—that does not require him to adopt a moral position and suggests that such facts ought instead to be evaluated according to pragmatic rather than ideological considerations: the Saracen language serves their needs, just as English serves "ours." The text does not at this point try to apply that conclusion retrospectively to the diversity of faiths, but the question of how the fact of difference ought to be evaluated unavoidably becomes a sort of hinge on which the book now turns, as Sir John himself now turns, in the next chapter, from the Holy Land to yet more distant prospects.

> Now sith I haue told you beforn of the Holy Lond and of that contree abouten and of many weyes for to go to that lond and to the Mount Synay and of Babyloyne the More and the Less and to other places that I have spoken [of] beforn; now is tyme yif it like you for to telle you of the marches and iles and dyuerse bestes and of dyuerse folk beyond theise marches. For in tho contrees beyonden ben many diuerse contrees and many grete kyngdomes that ben departed by the iiii. flodes that comen from Paradys Terrestre. (16.105)

But the emphasis on difference and diversity ("dyuerse bestes," "dyuerse folk," "diuerse contrees") has already been thoroughly undermined; we should not be surprised to find that the farther East, like the nearer East, will turn out to be home to much that is morally and politically congenial to Sir John's turn of mind.[27]

Cathay: Ethical Economies of Scale

Sir John's admiring account of the Saracens contains a powerful hint of what will turn out to be its sequel in the *Travels*. Distinguishing between the

Sultan's Babylon (i.e., Cairo) and the "gret Babiloyne" of biblical fame (i.e., Babel), he notes that the latter, further to the east, "is in the power and the lordschipe of Persye.

> But he holdeth it of the Grete Chane, that is the gretteste emperour and the most souereyn lord of alle the parties beyonde, and he is lord of the iles of Cathay and of manye other iles and of a gret partie of Inde. . .And he holt so moche lond that he knoweth not the ende, and he is more myghty and gretter lord withouten comparasoun than is the Soudan. Of his ryalle estate and of his myght I schalle speke more plenerly when I schalle speke of the lond and of the contree of Ynde. (6.29)

Sir John promises that, like all good sequels, his will be bigger, grander, and more spectacular by far than its original. Cathay thus represents, most fundamentally, an increase in scale over what comes before, and that increase presents both new problems—can Sir John's empirical claims really compel the reader to believe in such astonishing imperial magnificence?—and new opportunities to extend, historically as well as geographically, a coherent account of the world in which pagan virtue is not only a contemporary phenomenon but also the present result of recoverable historical circumstances. The Sultan lives in a virtuous present, in the *Travels'* idealizing account, but the Great Chan's virtue has its own legitimizing genealogy.

What the Chan has first, though, is a kingdom, and it is a realm so vast and wondrous that describing it sorely taxes even Sir John's rich vocabulary of superlatives. Indeed, its description can barely be contained within the text: from the first mention of the Chan in Chapter 6 (quoted above), his name appears regularly until the last chapter of the book (Chapter 34), and the description of his court and his vassals occupies six full chapters. His sway extends from history to myth: his marches border the land route to Jerusalem in the West, while in the East lie Prester John's land and the Earthly Paradise. Mandeville claims that "his lond and his lordschipe dureth so ferr that a man may not gon from on hed to another, nouther be see ne londe, the space of vii. yeer" (25.175). His royal progresses and the ceremonies of his court are impossibly extravagant, and his power is both the measure and summit of temporal authority. He is "the most gret empcrour that is vnder the firmament, outher beyonde the see or on this half" (21.139), "the gretteste emperour and the most souereyn lord of alle the parties beyonde" (6.29), and "the most myghty emperour of the world and the grettest lord vnder the firmament" (24.166). His kingdom contains the city of "Iyonge [Beijing]. . .that is a gret del more than Rome" (24.166), and twelve provinces ruled by twelve principal kings, "and euery of tho kynges han many kynges vnder hem, and alle thei ben obeyssant to the Gret Chane" (25.175).

The Chan's kingdom is thus a microcosm—though apparently just barely—of the world at large. What this means is that the religious diversity that has characterized the world so far—the world of Jacobites, Nestorians, Georgians and Greeks as well as Saracens and Western European Christians—is fully represented within the Chan's realm itself. Toleration for all creeds is the rule, though, and reassuringly for Mandeville's readers Christianity is at least first among equals. Although the Chan himself is not Christian, "natheles he wol gladly here speke of God. And he suffreth wel that Cristene men dwelle in his lordschipe, and that men of his feyth ben made Cristene men yif thei wile thorghout all his contree, for he defendeth no man to holde no lawe other than him lyketh" (25.176). The Chan goes well beyond toleration; he is downright reverent toward the cross, to which he respectfully kneels whenever he encounters a Christian settlement along his route. Further, he employs Christian physicians, whom he trusts more than he does his Saracen ones, and in his court there dwell "many barouns as seruytoures that ben Cristene and conuerted to gode feyth be the prechinge of religiouse Cristene men that dwellen with him" (25.172).

Mandeville's representation of the Great Chan as a Christian sympathizer recalls his efforts in the Saracen chapters to represent simultaneously the spiritual hegemony and historical crisis of Western Christendom, a faith wholly convinced of its possession of religious truth and wholly incapable of imposing that vision on the larger world, save through fictions like the *Travels*. Like the Saracen prophecies predicting the eventual though historically unspecified fall of Islam, the respected but hardly dominant place of Christianity in Cathay is the result of Mandeville's efforts to reconcile his religious orthodoxy with a fidelity to history, and with the fruits of his own putative observations. It is possible to read this attempted reconciliation hopefully, as holding out the possibility that the open-minded Chan might imitate Constantine and legislate the conversion of an entire empire.[28] More likely, though—or at least easier to predict—is a continuation of the status quo, since for all the Christians in the Chan's court "ther ben manye mo that wil not that men knowen that thei ben Cristene" (25.172)—a statement that recalls the Christians converting to Islam in Chapter 15 and suggests that the moral shortcomings of Christians in the Mediterranean are matched by a failure of nerve in Cathay. What renders this situation yet more poignant is the fact that, in the past, there have been Christian Chans. In his brief history of the Mongol empire in Chapter 24, Mandeville relates how "Mango Chan, that was a gode Cristene man and baptyzed, and yaf lettres of perpetuelle pes to alle Cristene men. . .sent his brother Halaon with a gret multytude of folk for to wynnen the Holy Lond and for to put it into Cristene mennes hondes and for to destroye Machametes lawe. . ." (24.165). Halaon (Hulagu to historians) according to Mandeville captured "alle the Lond of Promyssioun" and turned it over to Christian control. Mango's death, says Sir John, "was gret sorwe and loss to alle Cristene men" (24.166).[29]

At the same time this description of Hulagu's expedition against the Saracens does important work in domesticating the empire-building Tartars. Its evocation of authentic if short-lived thirteenth-century hopes for a Saracen-crushing alliance between Christians and Tartars helps to repress equally authentic thirteenth-century anxieties about unstoppable Tartar expansion. The Tartars may not be our coreligionists, but they are our friends, and they certainly aren't poised to besiege Vienna. Besides, is not toleration of all creeds (except, apparently, Islam, though historically the Persian Ilkhans were Muslims at this time) a hallmark of Cathayan civilization?

It is, although exactly why it is also turns out to be somewhat mysterious. Recalling again in Chapter 26 the Tartars' reputed thirst for conquest—the same trait that was put at the service of "alle Cristene men" by Mango and Hulagu—Sir John describes how they devote

> [a]lle here lust and alle hire ymaginacioun. . .for to putten alle londes vnder hire subieccoun.
>
> And thei seyn that thei knowen wel be hire prophecyes that thei schulle ben ouercome by archieres and be strengthe of hem. But thei knowe not of what nacoun ne of what lawe thei schulle ben offe that schulle ouercomen hem, and therfore thei suffren that folk of alle lawes may peysibely dwellen amonges hem. (26.181)

Here the toleration described in the previous chapter—"he defendeth no man to holde no lawe other than he lyketh"—is shown to be the result of a Tartar prophecy. But the "therfore" that introduces that custom implies a logical connection between the prophecy and the Tartars' accommodation to it that simply does not exist. The logical response to such a prediction would doubtless be pervasive xenophobia and a fundamental distrust of all things foreign, and in the source of the passage the Tartars clearly impose their law and customs on the peoples they conquer.[30] But the author of the *Travels* has already committed himself to a Cathay that embraces religious diversity, and he thus changes the passage to correspond to that vision. Moreover, by adding the detail of the archers, evidently his own invention, he accomplishes two things: on the one hand he adds another authenticating, Anglicizing note, since the successful use of archers figured prominently in English military strategy and mythology in the second half of the fourteenth century; on the other he makes the Tartar prophecy further resemble the earlier self-limiting Saracen prophecies of the fall of Islam. His source here is quite specific: the Tartars have eighteen years of rule left before their fall. Characteristically the *Travels* omits this detail, leaving the date historically obscure and the "nacoun" a little less so.

Most important for the *Travels'* representation of pagan virtue is the way in which the prophecy and the principle of toleration that is unaccountably derived from it serve formally (and circularly) to authorize such toleration as the proper response to cultural and religious diversity. The Chan's holdings are unimaginably vast and diverse, and as Sir John describes them they are apparently free of internal religious conflicts. The Chan is the greatest of all earthly rulers, a paragon of seigneurial virtues whose success in rule would be the wish of all the lords of the West (were its achievement not so fantastically unlikely), and one of his chief customs is a toleration that seems to contribute directly to the stability of his realm. While admittedly such a portrayal permits Mandeville to exalt the state of Christianity in the East by dwelling on the Chan's special affection for it, nevertheless its chief effect is a subtle but potentially profound levelling of all faiths. Faith and its expression are represented implicitly as just one more cultural practice to be catalogued and commented upon. We have already learned, in the Egypt section, that faith is not an essential measure of virtue, or rather that embracing any particular faith—say, Christianity—does not necessarily lead to the virtuous practice of it. Now, in Cathay, we learn that religious practice can be understood, in these circumstances anyway, as a subcategory of civic or state-oriented activity rather than the essential marker of a culture's identity.

These ought to be anxiety-provoking thoughts for Sir John, and they are. He displaces that anxiety, however, in keeping with his usual practice. The ultimate conclusion of these reflections—that God himself may weigh all faiths equally—he puts off to the end of the book, to the Brahmans and to the epilogue. In the Cathay chapters the anxiety manifests itself in a passage about the very nature of belief, one that reaches once again to the premises of Sir John's empirically based report. Isn't the real problem with Mandeville's story of Cathay the fact that it is simply unbelievable?

In an oft-quoted asseveration, he acknowledges his readers' potential skepticism and attempts to overcome it.

> wee hadden gret lust to see his noblesse and the estat of his court and alle his gouernance, to wit yif it were such as we herde seye that it was. And treuly we fond it more noble and more excellent and ricchere and more merueyllous than euer we herde speke offe, in so muche that we wolde neuer han leved it, had wee not a seen it. For I trowe that no man wolde beleve the noblesse, the richesse, ne the multytude of folk that ben in his court but he had seen it. (23.158)

Being there has helped Sir John overcome his own disbelief, but his assertions actually create more of a problem than they solve. His readers haven't been there, of course—what compels them to believe? Realizing that he has merely reinforced their skepticism, he tries to lull it logically into acceptance by providing some mundane particulars, again the kind only available to an eyewitness. Although the Great Chan feeds uncounted multitudes of retainers in his court everyday, the cost is really quite reasonable:

> But the ordynance ne the expenses in mete and drink ne the honestee ne the clennesse is not so arrayed there as it is here. For alle the comouns there eten withouten cloth vpon here knees, and thei eten alle maner of flessch and a litylle of bred, and after mete thei wypen here hondes vpon here skyrtes, and thei eten not but ones in a day. (23.158–59)

But this is just another unhelpful paradox, the notion that a resistance to luxury (and to good manners) makes possible a fabulous magnificence. Deciding perhaps that the savings in vegetables, napkins, and second helpings might not make the economics of the situation any more plausible, Mandeville simply refuses to be diverted by the Doubting Thomases in his audience any longer: "And whoso that wole may leve me yif he wille. An whoso wille not may leue also. . .And therfore I wille not spare—for hem that knowe not ne beleue not but that thei seen—for to telle you a partie of him and of his estate. . .(23.159)."[31]

"Beth war of men, and herkneth what I seye!" Sir John has reached Chaucer's Troilian impasse, armed with less than a full portion of Chaucerian irony. It's a fish-or-cut-bait moment in the *Travels* that broaches the theoretical limit of the entire empirical project: Sir John's belief (in anything—Cathay's splendor, miraculous relics, a sea of gravel) is based on observation, he claims, but ours is based on accepting that observation as valid, accepting it second-hand. And the answer to the question "Why should we?" seems to be "Why not? We have, more or less, for the first twenty-two chapters." Since (like any other author) he can provide no further theoretical justification for our accepting his authority, he falls back on assertion, and on habit, the habit of reading that has been cultivated in us through our experience of the text so far. This moment in the *Travels* recalls Sir John's colloquy with the Sultan, another place where the empirical premise of the book itself rises irresistibly to the surface. And like that earlier scene, it is immediately connected to some disturbing, or at least destabilizing, information about a pagan culture.

Immediately after the "Doubting Thomas" paragraph, the next chapter begins with a question of etymology. Why is he called the "Gret Chane"? The conventional wisdom, Sir John suggests, connects him with "Cham," that is, Ham, the cursed son of Noah, and Mandeville accepts the genealogical connection: "the sothe is this: that Tartarynes and thei that duellen in the gret Asye, thei camen of Cham" (24.161). Of course, endorsing this identification means contradicting two other staples of the conventional wisdom, first the post-diluvian political geography that typically associated Ham with Africa, Shem with Asia, and Japheth with Asia Minor (and later Europe), and second the notion that Ham and his progeny were cursed by Noah and God.[32] The Great Chan and his kingdom hardly seem cursed. But Mandeville ignores both of these difficulties, because the point he wishes to make is just the opposite: the Great Chan was not cursed but blessed, or at least commissioned, by God.

"But the emperour of Chatay clepeth him not Cham but Can, and I schalle telle you how," says Sir John, showing us this time that one letter's difference can be a matter of some significance. Borrowing from the history of the Armenian aristocrat Hayton, Mandeville tells the story of the rise to power of Changuys (Genghis), which begins with the intervention of a mysterious, angelic White Knight speaking "on Godes behalue inmortalle," who names Changuys "Chan" and calls upon him to free the seven tribes of the Mongols from servitude. Rather than a cursed descendant of Noah, the first Great Chan is a type of Moses, and upon his rise he embarks on a legislative campaign equal parts Pentateuch and *comitatus*.

> And thanne he made many statutes and ordynances that thei clepen *Ysya Chan*. The first statute was that thei scholde beleeuen and obeyen God inmortalle that is allemyghty, that wolde casten hem out of seruage and at alle tymes clepen to him for help in tyme of nede. The tother statute was that alle maner of men that myghte beren armes scholden ben nombred; and to euery x. scholde ben a mayster, and to euery c. a mayster, and to euery m. a mayster, and to euery x^m. a mayster.
> After he commanded to the princypalles of the vii. lynages that thei scholden leuen and forsaken alle that thei hadden in godes and heritage and fro thensforth to holden hem payd of that that he wolde yeue hem of his grace. And thei diden so anon. After he commaunded to the princypales of the vii. lynages that euery of hem scholde brynge his eldest sone before him and with here owne handes smyten of here hedes withouten taryenge. And anon his commandement was performed. (24.162)[33]

The subversive analogies in this passage—the Chan as God/Moses/general, his vassals as Israelites/Abraham/army—suggest the scale according to which Mandeville wants us to appreciate the Chan's authority, and at the same time point once again to the sort of rhetorical and spiritual hegemony of Christian values (here figured as specifically Old Testament ones[34]) that permeates the description of the Chan's kingdom. Indeed, at one point the two line up to produce a sort of inverted Exodus. On his path of conquest, Changuys heads for the territory beyond "the mount Belyan," and the

White Knight appears to advise him that he will find the sea impassable there unless he kneels nine times to the east "in the worschipe of God inmortalle" (24.164). Following these instructions, the Chan sees the waters recede to leave a path nine feet wide (since which time the Cathayans have held "the nombre of ix. in gret reuerence"). Although he is not a Christian, the Great Chan (like Alexander the Great, as we shall see) can have his prayers answered.[35] And they are portrayed as *his* prayers, offered according to divine wishes. The Chan's conquests receive not only divine approbation, like the depredations of some run-of-the-mill scourge of God, but divine commission. The Old Testament echoes suggest that the Tartars are in some way another chosen people (and thus these chapters too contribute to the *Travels'* ongoing displacement of the Jews).

Thus the alternate etymology Sir John offers for "Chan" carries along with it an alternate history for the nation of Cathay, and the beneficent intervention of the White Knight contrasts markedly with the curse known to lie on the descendants of "Cham." Sir John's version rewrites Cathay's role in providential history, basing it on a series of revelations rather than a curse—indeed, a recent series of revelations, since he says that a little more than eight score years ago the Tartars were still a nation enslaved (24.161). Mandeville's rejection of the conventional philological determinism of his age, underwritten as usual by the formal premise of his own personal ethnographic investigation, once again permits him to redefine the pagan world vis-à-vis his own.

"the yle of Bragman": *plus ça change. . .*

Mandeville's reworking of the Great Chan's place in the providential scheme of things can be interpreted as an attempt to make Cathay less of an anomaly, not more of one. Though the reciprocal relationship of political and moral authority first outlined in the Egypt chapters certainly still obtains in Cathay, Mandeville's hyperbolic presentation of the Great Chan's realm tends to muddle the distinction between different kinds of authority in the name of magnificence. The Chan's status as the world's greatest ruler tends to render all questions about authority rhetorical and self-referential— things are the way they are in his realm because he is who he is. The intervention of the White Knight, however, while it certainly provides Cathay with an unprecedented providential importance, also makes it something other than a *sui generis*, self-generated state. Its magnificence is thus connected, if only implicitly, to the divine favor shown to Changuys, and thus made intelligible to Western analysis. As with the Saracens' prophecies, Mandeville's efforts to demonstrate the existence of pagan virtue in unconventional places require him to employ some conventional devices.

He does, though, describe one place where Western readers did conventionally expect to hear tales of virtuous pagans. That place is the land of the Brahmans, where moral rectitude not only purchases peaceful stability, but guarantees meteorological advantages as well: "And because thei ben so trewe and so rightfulle and so fulle of alle gode condicouns, thei weren neuere greued with tempestes ne with thonder ne with leyt ne with hayl ne with pestylence ne with werre ne with hunger ne with non other tribulacioun as wee ben many tymes amonges vs for oure synnes" (32.212). With the Brahmans Mandeville reaches what was traditionally regarded as righteous heathen territory, and his liberal understanding of the issue manifests itself here most strikingly. Having already demonstrated an admiration for "trouthe" to one's "lawe," Sir John in this chapter rhapsodizes vigorously on that most delicate of subjects, the salvation of the heathen. But he also constructs his account of the virtuous Indians so as to recall, lexically and structurally, his earlier encounters with non-Christian peoples. The most powerful convention at play in this chapter of the *Travels* is precisely the convention of the righteous heathen, and by linking this section to earlier ones he mobilizes the ecumenical spirit of that theme on behalf of those peoples, who could not be successfully or openly associated with it in the earlier stages of the book. Conversion, for example, which is the ostensible goal of Sir John's analysis of the Saracens, does not arise as an issue in the Bragman chapter. If anything, Sir John suggests, *we* need to be converted—we need to become more like them.

Mandeville's description of the Brahmans and Gymnosophists derives ultimately from the widely circulated and much mediated romance histories of Alexander the Great, which I discuss in chapter 3. He depicts them in a manner perfectly in accord with received medieval opinion, as sober, chaste, simple, impossibly virtuous and nearly smug about it. Their ethical rectitude, based on adherence to natural law, translates into perfect social stability, and the strictness of the adherence recalls the emphasis on "trouthe" and "deuocoun" so celebrated by Sir John in earlier chapters. For do not some call the "yle of Bragman" the "lond of feyth?"

alle be it that thei ben not cristned ne haue no perfyt lawe, yit natheles of kyndeley law thei ben fulle of alle vertue. And thei eschewen alle vices and alle malices and alle synnes, for thei ben not proude, ne coueytous, ne envyous, ne wrathfulle, ne glotouns, ne leccherous, ne thei don to no man otherwise than thei wolde that other men diden to hem. And in this poynt thei fulle-fillen the x. commandementes of God, and yif no charge of aveer ne of ricches. And thei lye not ne thei swere not for non occasioun, but thei seyn symply *ye* and *nay*. For thei seyn he that swereth wil disceyue his neyghbore, and therefore alle that thei don thei don it withouten oth. (32.211)

Here again is the spiritual and rhetorical hegemony of the Christian
faith, in the Brahmans' fulfillment of the Ten Commandments according to
"kyndeley law," with the additional amplification of the Gospels' Golden
Rule and the avoidance of six out of seven deadly sins. The Brahmans
"leden so gode lif as that thei weren religious men. . .Wherefor it semeth
wel that God loueth hem and is plesed with hire creance for hire gode
dedes. Thei beleven wel in God that made alle thinges, and Him thei
worschipen" (32.212).[36]

They worship the right God, as do the Saracens, and they seem to
receive his approval, as the Cathayans have. But Sir John presses yet further,
in an original passage that puts him squarely on the liberal side of contemporary thinking about the righteous heathen:

> And alle be it that theyse folk han not the articles of oure feyth as wee han,
> nathelees for hire gode feyth naturelle and for hire gode entent I trowe fully
> that God loueth hem and that God taketh hire seruyse to gree, right as he did
> of Iob that was a paynem and held him for his trewe seruant. And therefore
> alle be it that there ben many dyuerse lawes in the world, yit I trowe that God
> loueth alweys hem that louen Him and seruen Him mekeley in trouthe, and
> namely hem that dispysen the veyn glorie of this world, as this folk don and
> as Iob did also. (32.214)

The levelling of faiths hinted at in the Cathay chapters here springs forth
fully accomplished, and Sir John goes on for two more lively paragraphs,
adducing Scriptural support for his assertions, fideistically defending his
position ("for wee knowe not whom God loueth ne whom God hateth")
and then immediately turning around excitedly to overwhelm his own
defense ("And therfore seye I of this folk that ben so trewe and so feythfulle
that God loueth hem. . .").

But if Mandeville reaches rhetorical heights here because he has finally
found the ideal test case, the exemplary Brahmans whose legendary virtue
depends upon centuries of tradition, he does not let us forget any of his earlier cases, or even the case of Western Christendom itself. The intelligibility
of the Brahmans in the *Travels* depends upon those previous encounters,
and that dependence is made clear through two mirrored tales of Alexander,
describing his encounter with the Brahmans and his similar meeting with
the Gymnosophists. In condensing and reworking his sources, Mandeville
has the former accuse Alexander of intending to "disherite" them, a word
Mandeville had used in the prologue to describe the Western lords who
should be leading crusades but who prefer instead "to disherite here neygh-
bores." The Brahmans' claim that they have a king for the sake of inuring
obedience, not providing justice (for they are all thoroughly and individually

just) further recalls the common folk of the prologue, who can do nothing without the leadership of their temporal lords. Finally, the Brahmans' self-portrait describes how they eschew showy array, preferring "a sely litylle clout"; "Whan men peynen hem to arraye the body for to make it semen fayrere than God made it, thei don gret synne. . ." (32.212). This remark echoes the Sultan's claim that Christian folk "knowen not how to ben clothed," indulging vainly in all sorts of fashionable excess. Insofar as he acts the role of Sir John before the Sultan here, Alexander's reaction to the Brahmans' claims once more seems to reinforce the live-and-let-live pacifism endorsed in the earlier episode: he decides that to trouble them would be "gret synne," and he promises to leave them alone and in peace.

When he enters the land of the Gymnosophists "to see the manere"—recalling perhaps Sir John's motive for travelling to the Great Chan's court—Alexander shows himself to be a fairly quick study:"whan he saugh hire gret feyth and hire trouthe that was amonges hem, he seyde that he wolde not greuen hem, and bad hem aske of hym what thei wolde haue of him. . ." (32.213).Typically, they ask for immortality, a request that leaves Alexander "gretly astoneyed and abayst and alle confuse[d]. . . ." Just as in Egypt, in the Sultan's chamber, a confrontation between a representative Westerner (Mandeville/Alexander) and a representative Easterner (Sultan/Brahmans-Gymnosophists), placed at the end of a longer relation characterizing an unfamiliar culture, leaves the Westerner "gretly astoneyed and abayst and alle confuse[d]," deeply impressed and respectful of the alien culture and somewhat dubious about his own. This pattern—the virtuous pagan scene—represents a standard model for an East–West encounter in the *Travels*; it not only dramatizes the theme of the text as a whole, but structures the text as well.

At the same time the use of Alexander here in place of Sir John himself has several important effects. In the first place it removes the issue of conversion from the episode; Mandeville can observe from a reflective distance that the Brahmans "haue no perfyt lawe," but at the scene Alexander, himself a pagan, has no investment in leading them to any particular revelation, and this permits their manifold virtues to take center stage in the drama of their encounter. In addition, the use of Alexander the conqueror instead of Mandeville the ethnographer helps to reinforce one of the lessons learned in Cathay, that the appropriate response to diverse or unfamiliar religious practice is a respectful toleration of it. Thus Alexander, too, is an exemplary figure in this chapter; he may not possess the monastic virtues of the Brahmans, but he does recognize virtue when he encounters it, and he shows admirable restraint in allowing it to flourish unconquered. The chapter's theological aim—that is, its attempt to demonstrate (or at least assert) divine approval of the righteous heathen—is therefore paired with a secular one, its implicit valuation of a particular kind of worldly leadership. Alexander may be a stand-in for Sir John in his exchange with the Brahmans, but he is also another model for the temporal lords of the prologue, whose initial task—leading a crusade to retake the Holy Land—now seems not only more distant but also less necessary, less relevant to the world that the *Travels* has described.

In fact Alexander the Great pops up all along Mandeville's route as a sort of touchstone; Sir John virtually follows in his footsteps, from "Macedoigne, of the which Alisandre was kyng" (3.11), past the sites of his cities (most named after himself—"and yit he made xii. cytees of the same name," 17.114; see also 6.31, 7.33, 8.41, 27.185), his wars (against the king of "Chana," 18.120–21), and his miraculous deeds (enclosing the Lost Tribes, a famous medieval legend,[37] 29.192; the Trees of the Sun and the Moon, which prophesied his death, 32.215). The Gymnosophist episode is the last place he appears by name, but he is also a tacit presence in the following penultimate chapter, where Sir John discusses the route to the Earthly Paradise; the *Travels'* version of the approach to Paradise strongly echoes the Latin accounts of Alexander's attempt to reach it as described in the twelfth-century *Iter ad Paradisum*.[38] It lies beyond Prester John's lands, past mountains and "roches fulle grete," past a land of impenetrable darkness, behind an impassable wall and at the top of the highest mountain on earth. Even the indefatigable Sir John admits that he did not go there: "Of Paradys ne can I not speke propurly, for I was not there" (33.220).

> And yee schulle vnderstonde that no man that is mortelle ne may not approchen to that Paradys. For be londe no man may go...And be ryueres may no man go...Many grete lordes han assayed with gret wille manye tymes for to passen be tho ryueres toward Pardys with fulle grete companyes, but thei myght not speden in hire viage. And manye dyeden for werynesse of rowynge ayenst tho stronge wawes. And many of hem becamen blynde and many deve for the noyse of the water. And summe weren perisscht and loste withinne the wawes. So that no mortelle man may approche to that place withouten specyalle grace of God, so that of that place I can seye you no more. (33.221–22)

In the *Iter* Alexander and his host successfully reach their goal, though they are not permitted to enter; in the *Travels*, by contrast, neither Sir John nor any "grete lordes" are permitted to approach.

In fact, Mandeville's categorical declaration about the inaccessibility of Paradise is a crucial claim for the combined spiritual and geographical project of the *Travels*, for it defines by exclusion the physical world and the human community inhabiting it. His earlier "proof" of the earth's spherical shape and the possibility of circumnavigation, developed at great technical and anecdotal length in Chapter 20, had defined the world *inclusively*: you can get there from here, and when you get there—wherever there is—you will find "men, londes, and yles as wel as in this contree" (20.134).[39] That claim is specifically recalled in the Paradise chapter; describing how it can be dawn in Paradise but midnight in Europe, Sir John attributes the phenomenon to "the roundeness of the erthe, of the whiche I haue towched to you of before" (33.220). Geographically, the world is continuous and circumnavigable, and the one place that is inaccessible, the Earthly Paradise, is inaccessible to all, to Christians, Saracens, Cathayans, and even the virtuous Gymnosophists and ascetic Brahmans. This geography implies that all places are equidistant from God, and with that implies—or at this point virtually confirms—a sort of theology: all peoples are equidistant from God. Thus, all forms of mediation between humankind and God—all forms of worship—tend to be levelled, and the distinctions between creeds exist not so much in differences of doctrine, but in degrees of devotion. Intimations and brief sketches of this conclusion are of course scattered throughout the *Travels*, but it is really only at this point that all the separate pieces are drawn together, not just the thought-provoking accounts of non-Christian peoples but the related geographical principles of circumnavigation and Edenic exclusion. It is an especially witty and ironic touch that in a text so thoroughly given over to empirical observations, Sir John makes the final piece of his mappamundi puzzle depend on the one place he hasn't been to.

The Proof Is in the Papacy

The final chapter of the English texts of the *Travels* serves as a condensation of what has preceded it, a complex recapitulation of the relationship involving devout pagan customs, the salvation of the heathen, Mandeville's own Catholic orthodoxy, and the problem of authority in empirical reportage. That this is particularly the case in the Middle English manuscripts is due to the inclusion in all the English texts of the so-called papal interpolation, an episode not in the Insular French text, that recounts a fanciful detour to Rome, where Sir John submits his apparently already written book to the Pope for his imprimatur. This episode has been used in efforts to date the English versions of *Mandeville's Travels*, since placing the Pope in Rome as opposed to Avignon clearly says something—though it's not clear exactly what—about the date of the texts' composition. Within the text, though, the episode is clearly concerned with authority and orthodoxy, and it once again recalls earlier moments in the *Travels* where these issues appear.[40]

The final chapter describes three last "yles" before coming to its conclusion, and the most striking of these is the land of Ryboth, which like the other two is under the Great Chan's suzerainty. Ryboth is a "fule gode contree" in which no blood is ever shed out of reverence to a local idol, but the funeral customs of the Rybothans command most of Mandeville's attention: the priests feed the dead man's flesh to the birds and the son and kin devour the brain and make a goblet of the skull. The passage reverberates with overt echoes of Christian services: "as prestes amonges vs syngen for the dede *Subuenite sancti dei, et cetera*, right so tho prestes syngen"; "And of the brayne pan he leteth make a cuppe, and therof drynketh he and his other frendes also with gret deuocoun in remembrance of the holy man that the aungeles of God han eten" (34.225). Mandeville's source for this passage, Odoric of Pordenone, called this practice "vile and abominable," and later readers have certainly shared his reaction. Hamelius, the EETS editor of the *Travels*, found the scene a "grotesque burlesque," which is only heightened by reference to "the pope of hire lawe that thei clepen *Lobassy*. This Lobassy yeveth alle the benefices and alle othere dignytees. . .And alle tho that holden ony thing of hire chirches, religious and othere, obeyen to him as men don here to the Pope of Rome" (34.224).[41] On the opposite side of the spectrum, however, is Donald Howard, who points out how the practice of cannibalism, earlier repudiated (the inhabitants of Lamory engage in the "cursed custom" in Chapter 20), is here "impregnated with filial piety, with tenderness, with dignified family love, and redolent of the Holy Communion. . . ." The passage "asks us to behold a cannibalism not savage or repugnant but tender, dignified, and pious. It subtly reminds us of the Christian rites at which the Body and Blood of the Lord are consumed, but throws attention on the spirit in which men perform such rites."[42] Howard's reading clearly attends more closely to the spirit in which the passage is adapted from Odoric, though it is hard to imagine a medieval reader who would have found Sir John's evocation of Christian rites here at all subtle. But Howard's reference to an earlier account of cannibalism does point to an important structural connection between this episode and what has come before that can help us understand why the Rybothan custom is described not only as it is, but where it is, in the last chapter of the *Travels*.

In fact what has occurred is a small structural inversion: Chapter 20, in which the people of Lamory are condemned for cannibalizing children, goes on to include Mandeville's comprehensive defense of circumnavigation, while Chapter 34, in which the Rybothans are pointedly not criticized for cannibalizing their parents, actually begins with a remark about circumnavigation. From the antipodal "yles" around Prester John's land, Sir John says, one could return home via the circuitous route: "men myghte don it wel that myght ben of power to dresse him thereto, as I haue seyd you before," but most choose to turn around, and "returnen from tho yles abouesayd be other yles costynge fro the lond of Prestre Iohn" (34.223). One of those "other yles" is Ryboth, and thus it is a place Sir John encounters on his way back. His reference to the difficulty of circumnavigation and the choice to turn back here specifically recall the similar moment, in Chapter 20, "as I haue seyd you before":

And alle be it that it be possible thing that men may so envyroune alle the world, natheles of a m. persones on ne myghte not happen to returnen to his contree. For, for the gretness of the erthe and of the see, men mey go be a m. and a m. other weyes, that no man cowde redye him perfitely toward the parties that he cam fro but yif it were be aventure and happ or be the grace of God. (20.136)

But just prior to this passage, Mandeville has told a story precisely about circumnavigation taking place through "aventure or happ." "A worthi man departed from oure contrees for to go serche the world," he says, and after many years' travel eastward "he fond an yle where he hered speke his owne langage, callynge on oxen in the plowgh such wordes as men speken to bestes in his owne contree; whereof he hadde gret meruayle. . .(20.135)"— too much "meruayle," it turns out, for the astonished traveller immediately turns around to retrace his steps, journeying for a great while during which "he loste moche peynefulle labor."[43] Years later, an accident reveals to him what Sir John could have told him all along:

. . .that he had gon so longe be londe and be see that he had envyround alle the erthe, that was comen ayen envirounynge (that is to seye, goynge aboute) vnto his owne marches. And yif he wolde haue passed forth, he had founden his contree and his owne knouleche. (20.135)

We have seen this traveller's "gret meruayle" before, in the Sultan's chamber; it is the shock of the familiar encountered unexpectedly. But once Sir John has reached the limits of his travels and turned back, that shock has disappeared: the Rybothan funeral customs are reminiscent of Christian rituals, but not marvelously so, and Sir John's comparisons are contextualizing without being moralizing. The reference to circumnavigation with which the chapter begins, though it marks the end of Sir John's forward progress on his journey, also points to the context in which we can assess his progress in reconciling the various faiths and customs of the world, for circumnavigation is, as his previous chapter has shown, the geographical corollary of his concept of pagan virtue.

Mandeville's apparently final words on that concept occur a few paragraphs later.

And yee schalle vndirstonde that of alle theise contrees and of alle theise yles and alle the dyuerse folk that I haue spoken of before and of dyuerse lawes and of dyuerse beleeves that thei han, yit is there non of hem alle but that thei

> han sum resoun within hem and vndirstondynge—but yif it be the fewere—and
> that han certeyn articles of oure feith and summe gode poyntes of oure
> beleeve; and that thei beleeven in God that formede alle thynge and made the
> world and clepen Him God of Nature, after that the prophete seyth, *Et
> metuent eum omnes fines terre*, and also in another place, *Omnes gentes seruient ei*,
> that is to seyne, Alle folk schul seruen Him. (34.227)

Here Mandeville's ecumenism is more conservatively stated than in the
Brahman episode, though he does not unsay or retract anything he has said
so far. In fact, the emphasis on non-Christians sharing "certeyn articles
of oure feith" reinforces rather than repudiates his comparison of the
Rybothan prayers to *Subvenite sancti dei*. Moreover, the paragraph makes
clear in brief what the *Travels* overall has tended to show, which is that particularly in matters of faith diversity is often only superficial and that exotic
rituals only screen more fundamental similarities. This passage suggests that
the toleration shown to other cultures by the Chan and Alexander (and
Sir John himself) may really be an oblique, active form of self-recognition,
an encounter with one's "owne knouleche."

In the English manuscript tradition, these conclusions are approved by
the Pope himself, in an interpolated passage that cannily replays the themes
and scenes I have been discussing. On his return, Sir John arrives in Rome
with a desire to confess himself "of many a dyuerse greuous poynt" in his
conscience, "as men mosten nedes that ben in company dwellyng among so
many a dyuerse folk of dyuerse secte and of beleeve as I haue ben" (34.228).
The redactor (or interpolator) thus picks up on the "dyuerse folk" and
"dyuerse beleeues" of the just-concluded peroration on pagan virtue, and
implies that the Pope reads diversity in the same way Sir John does, as interesting but non-threatening. Sir John's diverse sins, attributed to his diverse
adventures, are apparently forgiven, and his exposure to (and of) the diversity
of the larger world is implictly held to be harmless.

Mandeville then submits his book, "this tretys," to be examined and
corrected.

> And oure holy fader of his special grace remytted my boke to ben examyned
> and preuyd be the avys of his seyd conseille, be the whiche my boke was
> preeued for trewe; in so moche that thei schewed me a boke that my boke
> was examynde by that comprehended fulle moche more be an hundred part,
> be the whiche the *Mappa Mundi* was made after. (34.229)

Iain Higgins aptly characterizes the circular procedure described here:
"[Sir John] offers his eyewitness testimony to the existence of a written
authority that testifies to the existence of what he as an eyewitness has
written about in his book—and this looping Roman dance is itself then

authenticated by an appeal to a partly written authority that readers might have been able to see for themselves: the *Mappamundi*, a pictorial and textual authority so well known that Sir John can point to it with a simple definite article."[44] This roundabout search for authentication and authority recalls Sir John's similar efforts when describing the court of the Great Chan—here he falls into a similar state of hyperbole ("moche more be an hundred part")—and as Higgins observes the final line of the interpolation echoes exactly that moment in Chapter 23:

> And so my boke, alle be it that many men ne list not to yeue credence to nothing but to that that thei seen with hire eye, ne be the auctour ne the persone neuer so trewe, is affermed and preued be oure holy fader in maner and form as I haue seyd. (34.229)

I dare you to disbelieve the Pope, Sir John is made to say, as the Cotton interpolator not so subtly links the text's theoretical problem of belief—that is, the grounds for accepting Sir John's empirical testimony—with the issue of right belief, the orthodox embrace of the spiritual authority of the papacy. Such a move would seem to change the terms of the question, though as we have seen the Pope himself has resort to a book to "prove" the truth of Mandeville's text; this passage also shows its author working hard to close one of the text's still open questions, using the materials and motifs already at hand.[45]

In fact, I would argue that the Cotton interpolator has another, still earlier moment in mind too. Sir John's encounter with the Pope recalls and partly inverts his earlier interview with the Sultan in Egypt, an echo confirmed by the use of the phrase "special grace": while in Egypt Sir John carried letters bearing the Sultan's great seal, "in the whiche lettres he commanded of his specyalle grace to alle his subgettes to lete me seen alle the places and to enfourme me pleynly alle the mysteries of euery place. . ." (11.60).[46] In his colloquy with the Sultan, Sir John had received some disconcerting information unasked for, while in his encounter with the Pope he makes a specific request for information to confirm his book; in Egypt he had sought to be shown all place and all mysteries, while in Rome he freely shows his conscience and his "tretys" to "his holy fadirhode." Indeed, Sir John has become the world he describes in his book, and both he and the book are here "preeued for trewe. . .alle be it that many men ne list not to yeue credence to nothing but to that that thei seen with hire eye." But in seeking to authorize and supersede the earlier interview, the papal interpolation is also necessarily constructed so as to resemble it, and in this passage the Pope and the Sultan (and the Chan) are themselves seen, through all their apparent diversity of sect and belief, to be fundamentally alike and equally authoritative in the way they relate to Sir John and in what they know of the world.

In adopting and adapting one of the *Travels'* most important structural motifs, the "authoritative interview"—a.k.a., the virtuous pagan scene—the English author of the papal interpolation reveals himself to be an especially keen reader of Mandeville, and one of the first to fall under the text's spell. He performs a kind of textual circumnavigation, and arrives at the end to find himself speaking Sir John's own language, constructing his episode in paradigmatic Mandevillean fashion. Moreover, he also betrays his prior acceptance of one of the text's chief premises, the legitimacy of pagan virtue, since the interpolation concerns itself not with the orthodoxy of the various "merueyles and customes" Sir John has allegedly witnessed but with whether he has represented them truthfully and accurately, with whether his book can be "preeued for trewe." The appeal to papal authority may indeed link questions of representation to the issue of right belief, but the interpolation nowhere suggests that right belief is incompatible with Sir John's earlier claim that "God loueth alweys hem that louen Him and seruen Him mekeley in trouthe," be they Christian or pagan. The interpolator's invention thus clears Sir John's conscience on this and "many a dyueresc greuous poynt" because his own conscience is already clear on the issue; pagan virtue is, for him as for the original author of the *Travels*, a foregone conclusion.

CHAPTER 3

THE MIDDLE ENGLISH ALEXANDER

Alexander Speaks

Mandeville's version of Alexander's encounter with the Gymnosophists goes like this: having been greatly impressed with their "gret feyth and hire trouthe," Alexander "bad hem aske of hym what that thei wold haue of him, ricchess or ony thing elles"; when they ask him to make them immortal, he admits perforce that such a gift is not in his power to give. The Gymnosophists—who clearly relish opportunities like this—immediately ask him

> whi he was so proud and so fierce and so besy for to putten alle the world under his subieccioun, "right as thou were a god and hast no terme of thi lif, neither day ne hour; and wylnest to haue alle the world at thi commande-ment, that schalle leve the withouten fayle or thou leve it. And right as it hath ben to other men before the, right so it schalle ben to othere after the. And from hens schaltow bere nothyng. But as thou were born naked, right so alle naked schalle thi body ben turned into erthe that thou were made of. Wherfore thou scholdest thenke and impresse it in thi mynde that nothing is inmortalle but only God that made alle thing." (32.213–14)

Overpowered by this lecture, Alexander "gretly astoneyed and abayst and alle confuse[d] departed from hem," thoroughly chastened. His silence underscores the apparent justice and poignancy of the Gymnosophists' remarks, and in the very next paragraph Mandeville launches into his famous panegyric on the virtuous heathen and their ultimate acceptability to God, further reinforcing the moral authority of the Gymnosophists to deliver this *memento mori* catechism to the proud and ambitious of the world.

But the episode doesn't always end this way. In most other Middle English versions of this anecdote, Alexander is anything but reticent in the face of the Gymnosophists' critique. In passus XVIII of *The Wars of Alexander*, they ask him "Quarto hiȝis þou fra half to halfe & all þis harme wirkis?" to which he promptly replies, "Sire, be my croune,

> Þe cause at I haue
> Is purly gods prouidens; predestayned it is before.
> Ȝe se wele seldom is þe see with himself turbild
> Bot with þire walowand windis. My will ware to riste,
> Bot anoþir gast & noȝt my gast þareof my gast lettis."
> (*Wars* 4189–94)

Having spoken his piece, Alexander departs, leaving the virtuous nudists "wemles" ("unharmed," but also "stainless" or "pure") to ponder the implications of his claim of divine commission.

In the Middle English alliterative poem *Alexander and Dindimus*, Alexander draws out those implications himself. Here the conqueror is even more loquacious in his self-defense, and turns his reliance on providential direction into a virtual political anthropology. When the Gymnosophists ask "How miȝt þu kepe þe of sckaþe with skile and with trouþe / Aȝeins ryht to bireve rengnus of kingus?" Alexander replies,

> Þorou þe grace of God I gete þat I haue. . . .
> Men seþ wel þat þe see seseþ and stinteþ
> But whan þe wind on þe watur þe wawus arereþ;
> So wolde I reste me raþe and ride ferþe,
> Never to gete more good no no gome derie,
> Bute as þe heie hevene goodus wiþ herteli þouhtus
> So awecchen my wit and my wil chaungen
> Þat I mai stinte no stounde stille in o place
> Þat I ne am temted ful tid to turn me þennus.
> And sin we wetin hur wil to worschen on erþe
> We mowe be soþliche isaid hur servauntus hende.
> Ȝif God sente every gome þat goþ upon molde
> Wordliche wisdam and wittus iliche,
> Betur miȝhte no burn be þan an oþur;
> Apere miȝht þe pore to parte wiþ þe riche.
> Þanne ferde þe worlde as a feld þat ful were of bestes,
> Whan everi lud liche wel lyvede upon erþe.
> For þat enchesoun God ches oþur chef kingus,
> Þat scholde maistrus be maad ovur mene peple;
> And me is markid to be most of alle oþure,
> For þi y chase to cheve as chaunce is me demed.[1]

In this passage the poet of *Alexander and Dindimus* amplifies his source, which while it contains the field metaphor—"Si omnes unius intelligentie fuissemus, totus hic mundus sicut ager unus fuisset"[2]—does not populate the field with beasts, nor elaborate as fully the divine right of Macedonian conquerors, as this passage does. The amplification, indeed the entire speech, foreshadows Alexander's approaching encounter with the virtuous Brahmans, whose placid and egalitarian way of life will seem beastly to him. And despite—or perhaps because of—the casual switching from a singular deity to plural ones and back again, the speech emphasizes the theme of providential endorsement that manifests itself repeatedly in the more philosophical moments of the Alexander romances. Like Mandeville's Great Chan, Alexander is divinely commissioned to conquer.

Of course, the vast body of medieval writing about the conqueror offered many different Alexanders for consumption, more than just "the scourge of God" or "the epitome of worldly vanity" available in the Gymnosophist episode. There was the scientific Alexander, who not only reported on the marvels of the East, Mandeville-style, but who was also recognized as a correspondent on diet, hygiene, and politics through his association with the vastly popular, pseudo-Aristotelian *Secretum secretorum* and the pseudoscientific texts affiliated with it. There was the historical Alexander, part of Western Christendom's pattern of *translatio imperii* and the subject of extended attention in popular encyclopedic histories like Vincent of Beauvais's *Speculum historiale* and Higden's *Polychronicon*.[3] And even the Alexander of the romances, Latin and vernacular, was a figure of multifarious adventures, marvelous, amorous, and chivalric. To distill from these extensive sources some essential medieval view of Alexander would be virtually impossible, and literally misleading.

At the same time several Middle English accounts of Alexander's exploits are particularly relevant to a consideration of vernacular representations of the righteous heathen, and in this chapter I consider the ways in which medieval English readers could consider Alexander the Great a virtuous pagan, if not "godus friend" (as the Brahmans will claim to be) then at least as God's "hende" servant. Certainly Alexander's historical situation—he died in 323 B.C.E.—rules out his participation in a conventional virtuous pagan scene, a lively dialogue between a pagan and a Christian. Though his ghost may haunt every Western *Fürstenspiegel* through the era of Machiavelli, Alexander's corpse is essentially silent, his afterlife innocent of miraculous resuscitations. Rather, it is in his own legendary lifetime that Alexander enacts the virtuous pagan paradigm, in a series of encounters with non-Christian folk in which more often than not he plays the Mandevillean role of the representative Westerner, at once virtuous pagan and shadow-Christian, as the tale of his divine commission suggests.

Obviously his dealings with various virtuous Indians fall into this category, and in the last section of this chapter I offer an extended account of the way that *Alexander and Dindimus* structures itself according to the concerns of the righteous heathen theme and assimilates the pre-Christian Alexander to a contemporary (that is, medieval) standard of aristocratic behavior. But it is also the case that Alexander's several encounters with the Jews can be read in terms of the virtuous pagan scene, in both structure and content. That is, not only does he engage in the dialogue about religious law and belief typical of the motif; Alexander also treats the Jews he meets exactly as late-medieval Christian readers and texts tended to treat them. When dealing with the devout and loyal inhabitants of Jerusalem, adherents to the Mosaic law that medieval Christians understood to be the "true" Jewish faith, Alexander is reverent and courteous, while in the legend of "Alexander's Gate" he treats the renegade Jews of the Lost Tribes, who have allegedly fallen away from the Mosaic covenant, with the scorn and suspicion typically associated with late-medieval anti-Semitism, perfectly recapitulating the distinction between "Scriptural" and "historical" Jews that medieval Christendom inherited from the Pauline and Augustinian traditions.[4] Thus in later Middle English texts Alexander is recuperated as a virtuous pagan in a dual fashion, as a pre-Christian exemplar of noble behavior whose *bona fides* are tested specifically against the virtues of righteous archetypes like the Brahmans, and as proto-Christian judge of Jewish devotion and depravity. (In fact, there is a third area of particular interest, the mirror-for-princes Alexander famous for his education by Aristotle; I take up this topic in chapter 4, in the context of Gower's use of Alexander in the *Confessio amantis*.)

"so commune"

Before turning to the relevant episodes, it is worth sketching out Alexander's place—his omnipresence—in the literary culture of later medieval England. Quoting Chaucer's Monk—"The storie of Alisaundre is so commune / That every wight that hath discrecioun / Hath herd somwhat or al of his fortune"[5]—has become something of a commonplace for critics trying to demonstrate the popularity and ubiquity of Alexander material in the Middle Ages, and indeed the Monk's remark captures both the wide range of audiences ("every wight that hath discrecioun") to which stories of Alexander were accessible and the wide repertory of tales ("somwhat or al") available to those readers. Alexander texts were certainly everywhere in late-medieval England, in part because earlier texts preserved in eleventh- and twelfth-century manuscripts were still around in substantial numbers. And even in the tenth century, in his preface to the text that would evolve to become the ubiquitous *Historia de preliis*, Archpresbyter

Leo of Naples had argued that the story of Alexander ought to appeal to readers of every kind:

> Good it is for all Christian men, prelates as well as vassals, secular as well as spiritual, to hear and learn of the fights and victories of the most eminent of the heathen folk, albeit pagan, who lived before Christ's coming, for it spurreth all to nobler deeds.[6]

Sometime around 1400 the chronicler Thomas Walsingham, in the preface to his revised version of the so-called *Compilation of St. Albans*, a twelfth-century prose history of Alexander probably composed at the Benedictine monastery there, declared that although the story will be most interesting to knights, nevertheless it

> will not be less useful to religious and to those in the cloister, since it produces in the process of its discourse not a little pleasure, and perhaps it will lift such persons out of their sloth and relieve their tedium. And since the reading of this history produces pleasure, it removes opportunities for wandering unprofitably and effectively destroys and slays them.[7]

The campaign against unprofitable wandering was evidently widespread, since extant monastic library catalogues and a review of surviving manuscripts of known provenance reveal Alexander texts in every kind of monastic or mendicant house and secular establishment.[8] We can also point to numerous individual clerical owners of manuscripts containing Alexander texts, from the anonymous "notarial personage of Durham" who compiled (or caused to be compiled) Corpus Christi College, Cambridge MS 450, a varied collection of letters, documents, and dictaminal treatises that includes some of Dindimus's letters to Alexander condensed from Vincent of Beauvais;[9] to the nearly anonymous William of Dunstable (Donestaple), a Benedictine monk of Ramsey abbey in Huntingdonshire who is listed in the mid-fourteenth-century catalogue as the donor of a "Hystoria Alexandri;"[10] to the almost known John of London, possibly an associate of Roger Bacon and a major donor of books to St. Augustine's Abbey in Canterbury, including a "Gesta Alexandri magni" and a medical and philosophical miscellany containing spurious letters of Alexander.[11] Still better known is John Erghome, the Austin friar (and later prior) of York and author of the commentary on the prophecies attributed to John of Bridlington; his 1372 gift of books to the convent at York included a "Gesta Alexandri magni libri X" (probably a copy of Walter of Chatillon's twelfth-century verse *Alexandreis*) and a copy of Alexander's letter to Aristotle on the marvels of India.[12]

Finally, we can point to an abbot and three bishops: Prior Nicholas Herford of Evesham Abbey left a "Bellum Trojanum, cum vita Alexandri in quaterno" and a "Secretum secretorum" to the abbey on his death in 1392; Robert Wivill, Bishop of Salisbury from 1330 to 1375, owned a book of histories that included the Julius Valerius *Epitome* and the *Epistola ad Aristotelem*; John Stafford, Bishop of Bath and Wells from 1425–43, Archbishop of Canterbury from 1443–52, and Chancellor 1432–50, probably owned Glasgow MS Hunter 84, a large collection of histories and travel literature (including Mandeville) in the early fifteenth century; and Geoffrey Hereford, Bishop of Kildare from 1449–64, owned BL MS Cotton Cleopatra D.v, which included the *Epitome* and the *Collatio cum Dindimo*.[13]

Turning to secular owners we typically find, as we might expect, French texts. Guy de Beauchamp's famous gift to Bordesely Abbey in 1305 contained "Un volum de le Enseignem[en]t Aristotle, enveiez au Roy Alisaundre," "Un Volum. del Romaunce d'Alisaundre, ove peintures," and "Un Volum del Romaunce des Mareschaus, e de Ferebras, de Alisaundre,"[14] while Thomas, Duke of Gloucester, Richard II's uncle, possessed not one but three copies of the Alexander romance, presumably in French.[15] One of these books has been traditionally but by no means certainly identified with Oxford, MS Bodley 264, the French Alexander romance that also contains the unique copy of *Alexander and Dindimus*; we do know that this manuscript probably belonged to Humphrey, Duke of Gloucester, in the fifteenth century, and that it was purchased in London in 1466 by Richard Woodville, Lord Rivers (and Edward IV's father-in-law).[16] Finally, on her way to her wedding with Henry VI in 1445, Margaret of Anjou received from her escort, the Earl of Shrewsbury, a richly decorated and illustrated collection of French romances, including texts about Charlemagne, Ogier the Dane, Guy of Warwick, and others; the Alexander romance appears first in the volume.[17] But "secular" does not necessarily mean "aristocratic" or "French" when it comes to interest in Alexander; William Walworth, Lord Mayor of London and Wat Tyler's slayer in 1381, was evidently interested in Alexander too, as was the Yorkshire antiquarian Robert Thornton, who copied the English *Prose Alexander* around 1440.[18] About the same time Thomas Dautree, a very well-read lawyer from York, left to his son John both his "librum de Gestis Trojanorum" and his "meliorem librum de Gestis Alexandri"; upon his own death in 1459 John bequeathed the Alexander text and his sword to one of his own sons, named, appropriately enough, Alexander.[19]

We could also include John Gower among the secular aficionados of Alexander literature; as we will see in chapter 4, he was as well acquainted with the range of texts available as any of his contemporaries. And Gower brings us to the list of fourteenth- and fifteenth-century writers concerned with Alexander material, a list that recrosses the line dividing clerical and secular and that includes Chaucer, Lydgate, Hoccleve (whose *Regement of Princes* draws extensively from the *Secretum secretorum*), and the anonymous authors of *Kyng Alisaunder, Alexander A, Alexander and Dindimus, The Wars of Alexander,* and the *Parlement of the Thre Ages.* To this we can add Higden, who drew on several sources for his chapters on Alexander; his two translators, John Trevisa and the anonymous fifteenth-century writer responsible for the translation in BL MS Harley 2261; and the fifteenth-century author-scribe of Worcester Cathedral MS F 172, which contains a unique (and uniquely bad) Middle English translation of Alexander's letter to Aristotle.[20]

English Alexander writing—that is, the insular production of unique texts, translations, and compilations—reaches back to the Anglo-Saxon version of Alexander's letter to Aristotle preserved in MS Cotton Vitellius A.xv. In the twelfth century two important Latin productions appeared, the *Compilation of St. Albans,* a hybrid prose history of Alexander that borrowed from Pompeius Trogus, Orosius, Josephus, Augustine, Bede, Isidore, and others, now extant in two thirteenth-century manuscripts,[21] and the *Parua recapitulatio,* a short text designed to supplement the accounts in the Julius Valerius *Epitome* of Alexander's birth, his visit to Jerusalem, and the struggles of his successors, the Diadochi.[22] This impulse to expand and correct Alexander texts that were encountered as complete but that were nevertheless perceived as somehow deficient achieved, in the fourteenth century, an impressive intertextual density. Exhibit A in this regard is of course Bodley 264, in which the English *Alexander and Dindimus* is supplied to remedy a missing "prossesse" of the French romance, although the Brahman episode was never actually part of the textual tradition of the French Alexander story to which it is appended. But there are plenty of less spectacular examples of this phenomenon of supplementation. Another appears in Cambridge University Library MS Ll.I.15, a fourteenth-century miscellany that contains the frequent pairing of the Julius Valerius *Epitome* and the *Epistola Alexandri ad Aristotelem.* In this manuscript, however, the *Epistola* breaks off after a few leaves, to be continued by the addition in a later hand of the paired stories of Alexander's celestial journey and his diving-bell descent to the bottom of the ocean, both of them borrowed from the J² recension of the *Historia de preliis* in order to give a thematically appropriate conclusion to a presumably faulty exemplar of the *Epistola.*[23] A third example is Walsingham's revised text of the *Compilation of St. Albans* referred to earlier, surviving in Oxford, Bodleian Library MS Douce 299, usually dated to about 1400; in his version, which includes occasional commentary of his own, Walsingham supplements his exemplar—itself a unique account—with material from Higden and John of Salisbury unavailable to the original twelfth-century compiler, including Higden's version of the conclusion of Alexander's encounter with Dindimus, in which the former abases himself before the Indian king.[24]

In surveying this material, I have deliberately blurred the line between the different languages of the extant texts, in part to underscore the impression that Middle English Alexander writing emerges from a rich, complex, and lively Anglo-Latin environment. But I also want to give as broad a picture as possible of the audience for these works, the monks, friars, abbots, priors, bishops, aristocrats, mayors, lawyers, clerks, and poets with whom we can securely associate the various texts. It is an audience that intersects in suggestive ways with the audience that Anne Middleton has described for *Piers Plowman*, a group that shares "a common social location, and a range of activities and interests. Whether laymen or ecclesiastics, their customary activities involve them in counsel, policy, education, administration, pastoral care—in those tasks and offices where spiritual and temporal governance meet." Middleton goes on to argue that *Piers* particularly engages the "practical historical imaginations" of such readers, a group for whom "penetrating to historical precedents and foundations of both temporal and spiritual imperatives is a habitual way of thinking, a means of resolution, and a source of deeply invested emotion"—capacities "also exercised by historical 'romance' in most of its major late-medieval English aspects."[25] Widespread consumption of Alexander texts in late-medieval England, not just by those who wrote and copied history but by those who lived it, reflects a keen appetite for exploring and understanding the sources of vernacular authority, that is, the validity of secular modes of experience, organization, bureaucracy, and rule. The audience's desire to hear tales from another era, "or þai were fourmed on fold or þaire fadirs oþir" (*Wars* 3) bespeaks not just a taste for marvels, but also a deep emotional investment in "the foundations of possession and authority," represented in "storial" form.[26]

Each of Alexander's virtuous pagan scenes—his meetings with virtuous Indians and both "good" and "bad" Jews—repay this investment, for each of them considers foundational moments. In India, Alexander's lordly exercise of power confronts in the Brahmans a species of ascetic behavior preemptively tagged as virtuous, thanks to centuries-old legend, but because the confrontation takes place well before the possibility of Christian conversion, mediation, or revelation, the encounter between worldliness and withdrawal can be seen as authentically exploratory and imaginatively open-ended. Both sides can be judged wanting, according to the Christian dispensation, and thus neither side has to be in the course of the fiction. With the Jews, of course, Alexander confronts medieval Christianity's own point of origin, at a moment in history before Augustine's doctrine of relative toleration had become either possible or necessary; that Alexander's judgments conform *avant le lettre* to this formulation argues implicitly for a historical imagination that included, in its conception of the prehistory of Christianity, both Hebrew patriarchs and Macedonian conquerors.

Alexander in Jerusalem: "Ay moʒt he lefe!"

The story of Alexander's visit to Jerusalem derives ultimately from Jewish sources, in particular Josephus,[27] and like so much else the episode made its way through the *Historia de preliis* into the English Alexander literature. The story does not begin promisingly: when Alexander sends to Jerusalem for supplies to help him sustain his siege of Tyre, Iaudas the "bischop," as a vassal of Alexander's foe Darius, has to refuse. Alexander, mightily displeased, vows to teach the inhabitants of Jerusalem a lesson as soon as he has the time, and a few hundred lines later, "he graythis him swyth / And ioynes him toward Ierusalem þe Iewis to distroy" (*Wars* 1576–77).

Deeply alarmed at this prospect, Iaudas leads the people of Jerusalem in a crash program of praying, offering and fasting in the hopes of enlisting divine aid against the approaching scourge. The plan works; an angel appears to the bishop in a dream, advising that Jerusalem be made an open city to Alexander. If welcomed with open arms and festive decorations, the conqueror will spare the town. Iaudas obeys, and when Alexander's forces arrive they find the city decked out as if "ane of þe seuyn heuyns" and all the inhabitants clad in milk-white clothes, crying "Ay moʒt he lefe! Ay moʒt he lefe!" to Alexander, "þe gretest & þe gloriosest þat euir god formed" (*Wars* 1728, 1734).[28] Alexander sinks to his knees at this sight, not (as his lieutenants think) in obeisance to the bishop, but in reverence to the Tetragrammaton that Iaudas wears inscribed on his mitre. Alexander has seen this symbol before, he says, and proceeds to recount a prophetic dream of his own, one that echoes the Great Chan's vision of the White Knight in *Mandeville's Travels*.

> For in þe marche of Messedone, me mynes on a tyme,
> Þat slike a segg in my slepe me sodanly aperid,
> Euyn in slike a similitude & þis same wedis,
> For all þe werd as þis wee wendis now attired.
> And þen I mused in my mynde how at I myʒt win
> Anothire an[g]ell of þe erth þat Aysy we call,
> And [he] me thret to be thra & for na þing turne,
> Bot tire me titely þareto & tristly to wende.
> (*Wars* 1748–55)

Alexander's conquest (particularly of Asia), like the conquest of the Mongols in *Mandeville's Travels*, bears the stamp of divine approval, and like

Changuys Alexander learns of his commission in a dream, a dream that unlike other portents in the poem does not need to be explicated by a "diuinour."[29]

Appropriately grateful for this divine attention, Alexander not only spares the city but makes offerings in the Temple "as þe lawe wald," honoring "oure lord" (*Wars* 1774–75).[30] And as in the Gymnosophist episode to come, Alexander offers a boon to those he has befriended: "Sire, quat þou will in þis werd to wild & to haue, / Noȝt bot aske at Alexsandire [anything of] reson, / And I sall grant, or I ga, with a gud will" (*Wars* 1793–95). But the familiar motif functions rather differently here; Alexander's offer elicits neither the Gymnosophists' moral hectoring nor the philosophical scorn of Diogenes, who in another famous anecdote responded to the king's offer with the request that Alexander step aside and stop blocking the light.[31] Iaudas asks for something Alexander can actually provide: freedom to retain their faith, and relief from tribute for seven years.[32] Alexander is only too happy to comply with this request for religious tolerance, largely because the Jews have just demonstrated a few lines above that their faith can accommodate him. For immediately prior to Alexander's offer, Iaudas

> Bringis out a brade buke & to þe berne reches
> Was plant full of prophasys playnely all ouire,
> Of þe doctrine of Daniell & of his dere sawis.
> Þe lord lokis on [a lefe] & [i]n a l[yne] fyndis
> How þe gomes out of Grece suld with þaire grete miȝtis
> Þe pupill out of Persye purely distroy;
> And þat he hopis sall be he, & hertly he ioyes.
>
> (*Wars* 1777–83)[33]

Alexander was favored by God: not only his dreams and actions, and those of the bishop of Jerusalem, but Holy Scripture itself confirms it. Once again, as in the *Travels*, the Judeo-Christian prophetic tradition proves inclusive enough to accommodate a pagan prince. Of course, the accommodation is historically distant in this case, and the interpretation of the prophecy had been fixed since the time of St. Jerome (whose allegorical reading is here "anticipated" by Alexander's own); it does not, as in the *Travels*, pretend to apply to the contemporary political situation or to predict the irresistible hegemony of the one true faith. But it does extend that hegemony backwards in time; history is essentially defenseless against this kind of recuperative desire, and in fact it is under the distinguished rubric of history—and not, as in *Mandeville's Travels*, the more suspect fictive umbrella of first-person ethnography—that this instance of domestication takes place. Alexander's excesses are balanced, in a sense, by his election. One critic, surveying the earliest versions of the visit to Jerusalem, has claimed that it represents an attempt by the ecclesiastical hierarchy (first Jewish, then Christian) that "neutralise et legitime, en la canalisant dans des formes orthodoxes," popular sentiment for the heroic and imperial cult of Alexander, and that by replacing "la priere a l'empereur par une priere pour l'empereur, de même, refusant a Alexandre la qualite de dieu, ils en font un disciple conscient de leur Dieu."[34] Although by the fourteenth century the audience for this episode had changed from "les foules" to the aristocratic and monastic consumers of historical romance, the Jerusalem set piece performs essentially the same function, legitimizing the admiration an account of Alexander's life invariably excites by investing him with the prestige of prophecy and by foregrounding the fact of his participation in sacred history.[35]

It is important to recognize, however, that the "formes orthodoxes" legible in this scene are indeed plural, for the episode not only depicts Alexander reverencing the one true God, but also Alexander honoring devout Jews who reverence the one true God, mirroring (or rather, anticipating) the way that medieval Christians would be encouraged to treat Scriptural Hebrews—Hebrews whose own role in providential history is here confirmed by their devout prayers and their belief in the "prophasys" of the "brade buke." Alexander is recuperated as a virtuous pagan here not only via his inclusion in Scripture, but also through his attitudinal resemblance to contemporary medieval Christian folk, with whom he shares a certain Mandevillean ability to discriminate between praiseworthy devotion among non-Christians and its deplorable opposite. That opposite presents itself to Alexander's judgment in the legend of the Lost Tribes.

The Lost Tribes: "Treuþe" and Its Consequences

The story of "Alexander's Gate" was widely known in the Middle Ages, and versions of the tale appear in Mandeville, in the *Wars of Alexander*, and the *Prose Alexander*.[36] It also appears in the pastoral literature, and it is to a pastoral version I turn here—a text in which Alexander and Trajan actually make a joint appearance as righteous heathens. In half the extant manuscripts of the *Northern Homily Cycle*, which circulated widely from the late 1380s through 1440, the two pagan emperors feature in exemplary stories attached to the gospel exegesis for the Nineteenth Sunday after Trinity. This aggressively orthodox pastoral work probably first appeared in the very early part of the fourteenth century, one of the later encyclopedic vernacular texts ultimately inspired by the Fourth Lateran Council and given special impetus in England by Archbishop Pecham's 1281 *Syllabus*; its chief goal, like that of the *Cursor Mundi* and Mannyng's *Handlyng Synne*, was the education of the laity (and perhaps the less learned parish clergy) in the elements of faith. It was copied, expanded, and modified throughout the century, and after 1380 may have been promoted as an orthodox response to Lollard translations.[37] It was, in other words, apparently a text directed at an audience rather different from the "possessioners" for whom the alliterative romances were composed, and a text designed to reinforce the authority of the ecclesiastical hierarchy rather than permit some imaginative free play about the foundation of authority in general. Clerks, its prologue states, possess "lar of Godes worde / (For he haves in him Godes horde / Of wisdom and of gastlic lare, / That he ne an noht for to spare, / Bot scheu it forthe til laued menne, / And thaim the wai til hevin kenne)."[38]

One could argue, though, that these audiences might sometimes intersect, or at least that they might have intersecting interests. One point of convergence was clearly *Piers Plowman*, and in fact an A-text of Langland's poem appears with the *Northern Homily Cycle* in the Vernon manuscript, which was probably completed sometime after 1384.[39] Another was the question of how to understand virtuous pagans, and here the sermon for the 19th Sunday after Trinity provides a useful example of how that understanding might compass both a popular, legendary sense of continuity with the pagan past and a pastoral, theological decorum of religious instruction—both of which were very interested in Alexander the Great. The gospel text is Matthew 9:1–8, which tells this story.[40]

> Crist he [Matthew] seide · schiped ouer a see
> And com in to · his owne citee
> And men brouhte bifore hym · a man
> Þat was wiþ þe palesye · tan.
> And sone · Whon crist heore treuþe sawe
> To hem he wolde · his mihtes schewe.
> He spac sone · ful myldely
> To him þat lay · in palesy.
> Mi sone he seide · þou leeue in me
> Þi synnes ben · forȝiue þe.
>
> (f. 210v, col. a)

The Jews (Vulgate: *scribis*) standing by are scandalized by what they consider Christ's blasphemy; he, aware of their "wikked þouhtus," declares that he spoke in order to make known the power of the Son of Man to forgive sins on earth. In order to prove the point, he proceeds to heal the sick man.

> Þen seide crist · to þat seke mon
> Bifore alle þe Jewes uchon
> Rys up he seide · al hol and strong
> Ber forþ þi bed · and hom gong.

Þe seke mon ros up · al hale
And felede bote · of his bale.
He tok his bed · and hom he ȝede
And crist Was I · blesset · for his god dede.
Ffor al þe folk · þat stod aboute
Sauh þis miracle · and hedde doute
And seide · God blesset mote you be
Þat hast ȝiue mon · so gret pouste.
Þis is þe strengþe · of ure gospel
As seint matheu · wol us tel.
                                    (f. 210v, col. a)

The explication that follows the gospel falls into three parts. The first is a
conventional allegorization, borrowed from "a clerk / Þat muchel con · of
gostly werk," of the major terms of the story: the sea is the world, the ship
is the Church, the city is man's soul (hence Christ's home), the palsy, deadly
sin that must be driven out by Christ. The second part is a fivefold exposition of the meaning of sickness—literal sickness—in the life of humankind,
and in the context of the Christian's relationship with God (illness makes us
meek, punishes our sin, increases our reward in heaven, etc.). Several
Scriptural and apocryphal stories are supplied to illustrate these claims: the
sister of Moses, stricken for "grucchying" against her brother; Job; Tobias;
Lazarus; the sufferings of Herod and the disgusting diseases visited upon
him after the slaughter of the Innocents. This last example is narrated with
considerable relish.

But it is the third portion of the lesson that is most striking. Here the
sermon returns to the concept of "treuþe" very briefly evoked in the gospel
verses, and works up a whole second lesson on the subject. Evidently
"treuþe" has the power to guarantee one's prayers.

     But ȝit Wol We loke · forþur mare
What may ben undurstonded þare
Þat vre gospel · Wol schawe
Hou Jhu · heore treuþe sawe
Þat brouhte · þis seke mon hym to.
Þis ilke Word · Wol I · vndo.
Vre gospel seiþ · þat Jhu crist
Þis mon of palesye · Warischt
Whon he sauh heore treuþe · and heor þouht
Þat þis seke mon · bifore him brouht.
Bi þis o Word · may We seo riht
Þat treuþe is · so much of miht
Þat ȝif mon bidde · a bone þerinne
And he beo out · of dedly synne

He may beo siker · to haue his bone
ȝif resun Wole · þat hit beo done.
And þeih he [a]ske · unskilfulli
ȝit Wol God haue · of him merci
And make him · on þat Wyse to Wit
Þat his askyng Is · unskilfulliche.
Þat mai ȝe seo Wel · bi a tale
Þat I riht now · telle schale.

(f. 211r col. a)

That the passage commends the "treuþe" of the friends rather than that
of the sick man himself renders impossible any allegorical reading of the
palsied man as a virtuous pagan, rescued by recourse to Christ; his illness
does not represent paganism or unbelief, but sin and "folye" of Christians.
But this reading is not entirely out of keeping with way "treuþe" functions
in Middle English accounts of pagan virtue, because "treuþe" here describes
the motives and state of mind of the intercessors, those who call attention
to the situation of those in need of God's grace. Thus, the tale that follows—
"hou Gregory preyde for a dampned man"—makes "treuþe" the property
of the Pope, not the pagan, and foregrounds not the heathen's righteousness
but the virtues of the man who seeks his salvation.[41] Langland's Imaginatif
would have found this story heartening, for it places the power to recuperate
virtuous pagans firmly in the hands of the Christian "þat in treuþe · is
studefast:" *facientibus quod in se est, Deus non denegat petitionem*. But the preacher
is not finished: in the final *exemplum*, he will guarantee to virtuous pagans
themselves the same claim on God's attention. And his exemplary pagan is
Alexander the Great.

The final example of "treuþe" is the story of Alexander's enclosure of
the Ten Lost Tribes in the Caspian Mountains, another legend with wide
currency in the later Middle Ages.[42] The Northern Homily Cycle version
differs from the one contained in *Mandeville's Travels* by providing a background to the encounter, an account of the Jews' exile based on the Hebrew
scriptures' pattern of rejection and reconciliation. After twenty-four lines of
condensed biblical history, Alexander once again inserts himself into the
picture, happening along (like Gregory) many years later.

Qd. alex. inclusit iudeos infra montes capseos.
   Þe king of Babiloynes · toun
Hedde mony Jewes · in his [bann don]
And flemed hem · in to Wildernes
Ffor heor synnes · and heor Wikkednes.
And aftur he lette · maken a cri
Þat non of hem Were · so hardi

To passe out · of þat Wildernes
Þat fer bi ȝonde · paynym is.
Þis Wildernesses · as Bok us telles
Is loken a boute · Wiþ heȝe hulles.
Mony Wynter · aftur þis fare
Þe kyng Alisaundre · com þare
Whon he hedde · al þis World conquered
And al folk Was · of him a fered.
Þis forseide Juwes · þat Wonede þare
Glosede þe kyng · Wiþ muchel fare
Þat he Wolde · ȝiue hem leue · to gange
Out of þat Wildurnesse · so strange.
　　He asked · as þe Bok . vs telles
Whi þei Weore sperred · in þe hulles?
A seriaunt · to þe kyng onswered
And seide þo Jewes · þere Weore isperred
Ffor þei so ofte · god for sok
And to false maumetes · tok
And evere · vn kuynde · þei han bene.
Þat haþ ben ofte · on hem i seene
Ffor God haþ don · for hem Wel mare
Þen euer for folk · dude he are
And euere þei don · vuel a geyn
And to heore lord · ben vn bayn.
Þerfore Wol he · as good riht is
Þat þei Wone · in Wildernis.
　　Whon þe kyng · þes Wordus herde
To þe Jewes · he onswerde
Siþen ȝe han · oure god forsaken
And to maumetes · ȝe han [ȝ]ow taken
I · schal fonde · to sperre ȝow mare
Now ar euere · I · hennes fare.
And a non he lette · calle Werkmen
And maade hem Worche · in cleiȝ and fen
Fforte stoppe · alle þe corneres
Bi twene þe valeys and þe hulles.
But Whon he sauh · þat Monnes trauayle
Mihte not · to such Werk · auayle
Ne monnes Werk mihte not folfulle
Þe kynges desyres · and his Wille
Þe kyng him self · knelede þare
And preyede God · Wiþ louely fare
Þat he schulde folfille · his longinge
And vche hul · to oþur bringe.
And a non · as he hedde prayed
Þe grete hulles · þat Wyde Were strayed

88

Weore so faste · to gedere fest
Þat Soþ · ne North · Est · ne West
May no þing passe · out of þat lond
But hit beo · þe ffoul fleoand.
    ȝit dwellen · mony Jewes þare
And so schul þei · euer mare
Til a ȝeyn · domes day.
Þen schul þei alle · þeonnes stray
In moni londes · schul þei Wende
And Cristendom · schul þei schende.
Ffor ouer al · þer þei go
Cristene folk · schul þei slo.
Þei schul leue · on Antecrist
And Wene þat he beo · Ihu crist.
Al þis tale · I · haue ow told
To make ow · in oure hertes hold
Þat al þat Cristene mon · Wol craue
In studefast trouþe · he may haue.
Þat me ȝe seo · bi þis kyng
Þat Wiþ good treuþe · gat his askyng
And ȝit Was he heþene mon.
And þerbi may We · seon vchon
Þat cristene mon · ouhte muche more
Gete · ȝif he in trouþe Wore
Þen saraȝin · þat trouweþ nouht
Þat Jhu Crist on Roode · haþ hym bouht.
Þerfore bihoueþ · cristene man
Þat Godus marke · haþ on him tan
Won hope · oute of his herte caste,
Þat hee beo · in treuþe st[ude]fast.
Ffor god may he · not elles paye
Ne no good preyere · may he praye.
God ȝiue us grace · so to trouwe
And so in rihtwys dedes · grouwe
Þat We may · at vre lives ende,
In to þe blisse · of heuene Wende Amen.
         (f. 211r col. b–c)

Alexander's "good treuþe" is evidently as valid as Gregory's, despite his being a heathen (as we are twice reminded). Once again, the *exemplum* is offered unproblematically as a historical incident illustrating the power of "treuþe." That Alexander's successful appeal for divine aid is presented as a spur to Christians, who ought "muche more / Gete. . ./ þen saraȝin," is a rhetorical flourish that does not diminish Alexander's merit; it recalls Archpresbyter Leo's description of the appeal of Alexander's story, which "spurreth all to nobler deeds."[43] If anything, it underscores the felt continuity between the pagan world and the Christian, since the identical virtue of "treuþe" is equally operable—and equally pleasing to God—in both spheres.

It is perfectly logical that this sense of continuity should cohere around Alexander, and it is telling that this Alexander legend in particular should serve as an example for "laued menne." The enclosure of the Ten Tribes/Gog and Magog is a story about boundaries and the policing of boundaries; Anderson argues that in drawing a line between the civilized and virtuous West and the dangerous foreign elements that lurk just outside, the tale tells "the story of the border in sublimated mythologized form." Scott Westrem goes even further in exploring that sublimation: "Many of Christian Europe's keenest anxieties—parricide, infanticide, cannibalism, anomie, nomadism, foreigners, peculiar languages (and character systems), inhabitants of the North, and religious dissidence—in various combinations underlie specific invocations of 'Gog/Magog'."[44] But the boundaries are apparently flexible—Alexander stands inside them—and the anxieties of orthodoxy apparently do not include paganism, since even a "saraȝin" can have his prayers answered. Once again, Alexander is enlisted to represent the interests of the West, not as the providential scourge of Christendom's enemies but as a conscious supplicant to "oure god." Not only is this virtuous pagan part of the imagined "West," his very deeds help to define it.

The tale's eschatological anti-Semitism, perfectly typical of an orthodox pastoral text like the *Northern Homily Cycle*, only confirms this reading. Alexander's reaction, a curious inversion of Pope Gregory's petition for Trajan (after learning of the moral state of the Jews, he prays to God not that they be released but that they remain exactly where they are), "neutralise et legitime" anti-Jewish sentiment both by placing its origins long ago in a region far, far away, and by connecting it in an apocalyptic sense with contemporary concerns—the enclosed tribes are still there, waiting for the (now imminent) advent of Antichrist. Embedded in this legend, of course, are supercessional anxieties integral to Jewish–Christian relations in the Middle Ages, a topic to which I return in the conclusion. At this point it is simply worth noting again the "pagan foundationalism" of the legend, and the way in which it, like the Jerusalem encounter, is modelled on the virtuous pagan scene: Alexander meets Others not of his faith, interrogates them, and passes judgment on their "lawe," or rather, their failure to adhere to that law with sufficient devotion—the kind of judgment that readers of *Mandeville's Travels* would recognize instantly.

Alexander and the Indians

In a sense we could say that the Alexander tales take the premise of *Mandeville's Travels* one step further: if we think of pagan virtue as an organizing premise of the *Travels*, which must over the course of the text be

assimilated to a reflexive Christian worldview (culminating in the validation
of the papal interpolation), then the Alexander romances take off from a
similar premise—Alexander's self-evident chivalric aristocratic virtues—
but, set as they are in the epoch preceding the Incarnation, do not exhibit
the same pressure to domesticate pagan religions via prophecy or imputa-
tions of implicit faith. Thus, they represent for contemporary writers a kind
of laboratory for exploring the nature of pagan virtue in what we might
call, admittedly with some exaggeration, its native environment. The legend
of the Lost Tribes, characterized as it is by an all-too-familiar anti-Jewishness,
is perhaps not the best example of this experimental impulse, although it
does suggest among other things the ways in which God found uses for secular aristocratic agency well before the Incarnation. The episode works
Alexander (and the King of Babylon before him) into sacred history by
demonstrating how the spirit of seigneurial enterprise he embodies paves
the way for Christian hegemony. When we turn to *Alexander and Dindimus*,
however, we find that the issue of supercession is bracketed, so that the
virtues (and vices) of two non-Christian peoples might be examined on
what seem to be their own terms.

Alexander and Dindimus is an 1139-line Middle English alliterative poem
largely devoted to the famous correspondence between Alexander the Great
and Dindimus, king of the Brahmans who live along the shores of the
Ganges. It comprises five letters, three by Alexander, who opens and closes the
exchange, and two by Dindimus, who replies to Alexander's request for an
account of the Brahmans' way of life first with a description of their virtuous
simplicity and then with an indictment of the Greeks, whom he depicts as
gluttonous, ambitious, idolatrous, slave-owning, hell-bound flatterers who
engage in murder, adultery, theft and absurd funeral practices that tend to the
corpse rather than the soul. Alexander's riposte is to deplore the Brahmans'
bestial, meager existence, which he attributes to bad habits rather than virtue,
and to assure them that in fact they're the ones going to hell. This vitupera-
tive exchange has—remarkably enough—typically been classified and almost
immediately afterward dismissed as a rather colorless debate poem, lacking
both the topical relevance of *Wynnere and Wastoure* and the adept control of
conventions of *The Parlement of the Thre Ages*. Skeat, inevitably one of the
poem's first editors, claims that "The author of this ingenious arrangement
strove rather for oratorical effect than sought to inculcate a lesson. . . It is
merely a question of seeing what can be said on both sides. There is nothing
else to be learnt from the story of it."[45] Critical assessments range from this
relative high point, echoed by Wells and Lumiansky in their respective editions
of the *Manual of Writings in Middle English*, to J.A.W. Bennett's assertion that it
is "a text of cultural rather than poetic interest" and Thorlac Turville-Petre's
characterization of the poem as "incompetent and tedious."[46]

Though one still disagrees with Skeat at one's peril, I would nevertheless contend that there is much to be learnt from *Alexander and Dindimus*, and although it may not be very good as poetry—and here it is pretty much impossible to disagree with the majority opinion—it is in telling ways a paradigmatic alliterative poem, an almost perfect example of the species: a generic hybrid surviving in a single copy that can be linked to both the learned Latin material that underlies what we used to call the "classical corpus" of Middle English alliterative poetry and to some of the concerns of *Piers Plowman*, particularly in its interest in the topic of the righteous heathen.[47]

As Ralph Hanna has recently observed, Middle English alliterative romances, particularly that subspecies of historical translation to which *Alexander and Dindimus* is usually assigned, are typically concerned "with the social disruptions potentially inherent in every exercise of lordship."[48] Unlike many of its cognates in this alliterative tradition, however—for example, *The Destruction of Troy* or the Alliterative *Morte Arthure*—*Alexander and Dindimus* problematizes the exercise of earthly authority not by showing its inevitably disastrous and self-destructive consequences, but by juxtaposing it with an alternate, ascetic mode, by exploring the foundation of and justifications for lordship and rule on the one hand and Brahmanic withdrawal from all operations of power and politics on the other—and, finally, by refraining in the course of the poem from making a categorical declaration on behalf of either side. Moreover, the poem's juxtaposition of apparently irreconcilable modes of living, rather than rendering it simply a formal and "oratorical" exercise, actually connects it to the most dynamic of alliterative poems, that is, to *Piers Plowman*. Though the question of which "lif" is most "louable" is posed early and often in the alliterative texts that come down to us—starting, of course, with *Winner and Waster*—Langland's intervention on this issue helped fundamentally redirect the framing of such a question for his more serious-minded followers (and readers) away from recreative exposition toward what Middleton has described as "the deep historicity of the salvific imagination of the self."[49] Thus, the philosophical exchange between historical figures that we find in *Alexander and Dindimus* can be at once thoroughly conventional—indeed, the Brahmans' place in the Western imagination, their reputation for ascetic probity, hardly changes between 400 and 1400—and in its late-medieval English context authentically exploratory: if the poem lacks the urgency of *Piers* (and it does), it shares Langland's seriousness. Alexander's desire "Of wide werkus to wite and wisdam lere" (216) and to discover "þe best lorus of life and lawus of wise"(224) is particularly intelligible, I would argue, in the context of (and as the historical analogue of) the same kinds of inquiries that repeatedly give structure to *Piers Plowman*: "Thenne Thouht in þat tyme sayde this wordes: / 'Whare Dowel and Dobet and Dobest ben in londe, / Here is oen wolde ywyte, yf Wyt couthe teche; / And what lyues they lyue and what law þei usen, / And what þey drede and doute, dere sire, telleth.' "[50]

One strategy the poem shares with *Piers* and several of its alliterative cousins is of course the principle of debate, an observation that typically invites more of those unflattering comparisons with *Winner and Waster*. But unlike *Piers*, the poem's disputations—while occasionally pedantic—are not scholastic in form, and *Alexander and Dindimus* is also unlike its other relatives in that the exchange between Alexander and the Brahman king is conducted entirely through letters. The two do not meet face-to-face in a dramatic showdown, as do, say, Alexander and Darius in *The Wars of Alexander*, an encounter that permits a very forceful expression of one of that poem's moral themes ("alle þe welth of þe werld worthis at þe last / To cayrayne & corupcion clene alltogidere" [*Wars* 3382–83]).[51] Partly this is a function of translation; the epistle is the formal genre of the poem's ultimate source, the *Collatio Alexandri cum Dindimo*, and the letter format was carried over into the *Historia de preliis*, the immediate source for *Alexander and Dindimus*.[52] But the English poet strenuously maintains the motif, even to the point of contradicting (like his source) the logic of his narrative. An exchange of letters is required because the river that separates Alexander's troops and the Brahmans is deemed impassable due to the "multi ypotami et scorpiones et cocodrilli"[53] that inhabit it, according to the *Historia de preliis*. The author of *Alexander and Dindimus*, perhaps realizing that Alexander has crossed a few rivers in his time, increases the aquatic population in order to make it a truly daunting barrier, adding dragons and adders and "oþure ille wormus" that render the river utterly impassable, except for eight weeks in July and August—high season, evidently, among the Brahmans (151–60).[54] This multiplication of perils is an attempt to disguise the fact that the epistolary conceit of the poem is rather gratuitously maintained than a matter of narrative necessity. What is necessary is some means to transport letters, and despite the threat of "harm of þe hound-fich þat hovede þerinne," (164) one of the Brahmans—those notorious isolationists—hops into a boat that he seems to have on hand and sails quickly to Alexander's side of the river.

> Of þe seggus þat he sai biȝonde þe side stronde
> He dide calle for to come to carpen him tille.
> Whan þei hurden is houp hastiliche aftur
> A lud to a litil boot lepus in haste,
> And raþe to þe riche king rowmus alone,
> And aftur of Alixandre askeþ his wille.
>
> (165–70)

Presumably the boat is only large enough to hold one message and one messenger, since Alexander specifically enjoins a written response to his first "sonde": "And we ȝou praien, sire prince, prestly me sende / Alle þe lorus of ȝour lif in lettrus aseled" (225–26).

"Lettrus aseled"—a phrase not in the poem's Latin source—are of course the means by which kings correspond, and one effect of treating the exchange of letters as the main action of the poem is Alexander's assimilation into a late-medieval model of kingship; he presides over the dispatch of letters on which he "settus his sel" (183), deploying not his troops but the instruments of diplomatic correspondence. Indeed, he even seems to maintain a diplomatic corps to interact with the native population; when the "lud" in the little boat reaches the Greek camp, Alexander details another "wel-langaged lud" of his own to "let þe king sone / Aspien ful spedeliche bi speche of þe land / In what kyþ were þey kyd and what it called were. . ." (171–73).[55] The poet's addition of both the material detail of the seal (repeated five times) and Alexander's concern for translation (also absent in the Latin[56]) seems to me to move the poem's inherited epistolary conceit beyond the usual anachronistic historicizing program of medieval romance, in which ancient Greeks or Trojans or Romans are represented as participating in recognizably medieval social or amatory or military practices, and toward the sort of "documentary poetics" described by Emily Steiner as operating in *Piers Plowman* and some of the *Piers* tradition. If Langland "used the written record—the clerkly activities that produced it, its ceremonial delivery, and its archival afterlife—to describe the penitential writing of salvation history,"[57] the author of *Alexander and Dindimus* here offers a slice of that writing's prehistory, a documentary embodiment of some of its themes stripped of the soteriological immediacy of Langland's poem that nevertheless appeals to contemporary readers—particularly the putative aristocratic readers of Bodley 264, the luxurious manuscript in which the poem survives—in some of the same ways.

Alexander and Dindimus is in addition an installment in Western Christendom's long, orientalizing, and largely ahistorical romance with the Brahmans, which extends from before Augustine (who condemned them) and Jerome (who praised them) through the seventeenth century.[58] But by the fourteenth century the Brahmans had become implicated in the ongoing discourse of the salvation of the heathen, as is clear both in *Mandeville's Travels* and in Dante, who takes up the virtuous pagan problem for the last time in the *Paradiso* with reference to ". . .un uom nasce alla riva / dell'Indo."[59] *Alexander and Dindimus* participates in this discourse as well, and often when it departs from its source—by amplifying or tendentiously translating the *Historia*'s Latin—it employs the Middle English lexicon of pagan virtue that we have seen operating in *Piers, St. Erkenwald*, and elsewhere.

Consistent with its refusal to offer definitive conclusions, it uses these contemporary terms to refer to both of the poem's principals, Alexander and Dindimus.[60]

The passages in which the poem differs significantly from its Latin source fall into three groups, two of which may be dealt with summarily. In the first place are those lines where the Latin has irresistibly suggested a Scriptural text to the author, who then works up a full-blown allusion for his own poem. So, for example, the natural wisdom of the Brahmans leads them to acknowledge that God created the world through his Word: "per Verbum istud mundum creavit et per hoc Verbum uiuunt omnia." In the hands of the English poet, this observation expands into an anachronistic allusion to John's gospel: "Godus worþliche Word, as we wel trowen, / Is sone soþliche of Man þat in Himsilf dwelleþ / By which molde is ymaad and man upon erþe / And al þat weihes in þis word scholde wiþ fare. . ." (615–18).[61] A second independent feature of the translator's practice is the addition of phrases or lines that serve to enhance either the Brahmans' over- all godliness or the putative monotheism of Alexander—a habit that might best be seen as a secondary effect of the virtuous pagan theme in the poem. So, for example, "Nos. . .Bragmanes" becomes "We bredde breþurne in God, Bragmanus pore" (287) in the Middle English; the Brahmans eschew gluttony not only for health reasons but because "To Godus pay is our peple in bettur point founde" (315) than those who overindulge, and they view bodily ornamentation and beautification as not just immoral but impious: those that "craven by craft comelokur seme. . .gaynsain hure Saviour þat hem so made / And ben aschamed of His schap and Schewen hem ellus" (414, 420–21). Alexander, for his part, condemns the Indians' idleness in refusing to take advantage of the world's bounty by claiming that they "schewe. . .to hur Schappere schame for his sondus, / Þat so schinden His schap þat He ȝou scheweþ here" (959–60).[62]

The influence of the contemporary righteous heathen theme is most visible in a third set of passages, beginning with the letters in which Alexander and Dindimus first size one another up. These passages establish a particular context for reading their exchange, and the poet's decision to dip into the virtuous pagan lexicon is an attempt to make both the Brahmans themselves and Alexander's interest in them intelligible to an audience already acquainted with the issue, I would argue, from texts like *Piers* and *Mandeville's Travels*. The encounter may be an historical one, but its textual environment includes contemporary works that touch on the righteous heathen question as well.

The poet begins by increasing Alexander's desire to communicate with the Brahmans; the Latin text attributes to him "desiderium magnum," but the English poet makes his desire of long duration as well as great urgency.

When the first Brahman identifies himself, Alexander replies, "Sertus. . .þi sawe me quemus, / Me haþ longe to ȝour land liked to wende; / Wiþ ȝou to carpe in þis kiþ covaitede y ȝorne; / For miche, ludus, of ȝour lif listned ich have" (177–80). His lifelong curiosity about the Brahmans in particular is perhaps a droll comment on the Christian West's own longstanding fascination with them, but it is also an interest the poem goes on to explain at greater length in Alexander's first letter to Dindimus, in which he asks if everything he has heard about the Brahmans is true.

> ȝif y wisdam or wit in ȝour werk finde,
> Þat God aloweþ ȝour lif and likeþ ȝour dedes,
> Y schal ȝour costomus, king, covaite to holde,
> And fonde sor bi miȝht ȝour fare to suwe;
> For fram þe ȝouthe of my ȝer ȝerned ich have
> Of wide werkus to wite and wisdam lere.
> We weren tauht in oure time and tendide lorus
> Of oure doctourus dere demed for wise,
> Þat non haþel undur hevene so holi is founde,
> Þat miht alegge any lak our lif to reprove.
> But for y, ludus, of ȝoure lif swich a los hurde,
> Þat we discorden of dede in many done þinguus,
> And þat ȝour doctours dere don ȝou to knowe
> Þe best lorus of lif and lawus of wise,
> And we ȝou praien, sire prince, prestly me sende
> Alle þe lorus of ȝour lif in lettres aseled. . .
>   (211–26)

Two things in this passage, indeed in the first four lines, are worthy of note. First, Alexander's desire to follow the Brahmans' law, should it prove better than his own, right away casts him as a potential virtuous pagan, by its evocation of the "facientibus quod in se est" commonplace. Alexander's words recall Imaginatif's paraphrase of the *facere* ethic in his discussion of Trajan's exceptional salvation:

> Ac truþe þat trespased neuer ne trauersed ayeins his lawe,
> But lyueþ as his lawe techeþ and leueþ þer be no bettre,
> And if þer were he wolde amende, and in swiche wille deieþ—
> Ne wolde neuer trewe god but [trewe] truþe were allowed.
>   (B 12.287–90)

Alexander certainly offers to amend, given the right reason—"Y schal ȝour costomus, king, covaite to holde"—and the whole passage turns on exactly that principle of discrimination that both animates and troubles Langland's poem: what sort of lives are "allowed," and about which ones should we "alegge any lak"? Equally noteworthy here is the standard Alexander intends to apply in assessing the Brahmans' claim to "wisdam or wit"; he proposes to determine whether or not "God aloweþ ȝoure lif and likeþ ȝoure dedes." This line, not part of the Latin source for this passage, seems to me to be fundamentally analogous to Langland's usage: the grammatical subject of the verb in both cases is God, and it is God's approval—not Alexander's—that is ostensibly at issue.[63]

In his reply, Dindimus picks right up on Alexander's potential as a righteous heathen candidate, and introduces two more terms familiar from Langland's account of Trajan, "rihtewisnesse" and "lawe."

> Bi þi message, man, þat þou to me sentest,
> Whan we sihen þi sonde wiþ þi sel prented,
> We kenden þi covaitise and þat þou, king, wilnest
> Þe rihtewisnesse wite þat to a weih longus.
> In þat alowe I þe, lud, þat þe lef were
> Þe beste lawe to lere and lorus of witte;
> For riht wisdam is worþ al þe world riche.
>
> (255–61)

"Rihtewisness" connotes a far more extensive kind of virtuous behavior than the Latin's "perfecta sapientia," and Alexander's desire to know "þe beste lawe" does Imaginatif's "bettre" law one better. Both protagonists seem to understand explicitly the imperative at the heart of the "facere" ethic, that one is obliged to follow whatever creed most perfectly conforms to one's most rational apprehension of the way the universe operates.

One effect of this language, here and elsewhere,[64] is a more level playing field for the debate; by casting *Alexander* as the virtuous pagan, anxious to learn the best law and live the best life, the poem distances itself from the less savory side of his reputation and makes him a more appropriate debating partner for the pacific Dindimus. Moreover, despite the general high regard shown in the medieval West for the virtuous simplicity of the Brahmans, the *Collatio* and its more faithful descendants necessarily present the Indians as part of *Alexander's* story; his eastward march is the historical narrative that provides the frame for their exchange.[65] In our text, then, Dindimus does not play a hectoring Lady Philosophy to Alexander's Boethius, but the Owl to his Nightingale, and the exchange that takes up the bulk of the poem thus becomes a debate between two virtuous pagans, the one a righteous heathen of long standing in the medieval imagination and the other an equally familiar character newly recruited for the role. And this equipoise renders the debate legitimately as well as rhetorically inconclusive, as neither interlocutor can appeal to any authority beyond the shadowy "Schappere" that they both recognize, but with whom neither can communicate directly. The historical setting of the poem's events is never described in detail, but we know it is before the Incarnation, before the establishment of the Christian covenant that binds the poem's readers, before the possibility of any salvation, much less exceptional ones. Alexander can claim, as he does at the end of the poem, that God has doomed the Brahmans to hell, but he can't call upon Scripture or prophecy to back up his assertion, though he does employ there yet another Langlandian keyword: "Þerefore, seggus, as y saide, for sake of ȝour dedus / Mede mowe ȝe of God in no manere fonge" (1122–23). Thus, paradoxically, a debate between two virtuous pagans produces a fully secularized—that is, putatively non-Christian—set of arguments about the merits of worldly action versus ascetic withdrawal. The poem's use of the virtuous pagan theme and the contemporary language of the righteous heathen issue are the formal ground of its seriousness, as it seeks to examine the underpinnings of world-liness and asceticism without making an automatic appeal to the Christian dispensation to resolve its debates, and without reflexively adopting the mor-alizing attitude exhibited by some contemporary historical texts.[66] In late-medieval England, for example, both Higden's *Polychronicon* and the *Eulogium Historiarum* contained versions of the story (derived ultimately from Vincent of Beauvais) in which Alexander, having "lefte of al pompe and pride" (Trevisa's words), comes to Dindimus to learn from him in person.[67]

Alexander and Dindimus avoids this temptation to turn a moral imperative into a narrative one—indeed, the poem definitively rejects such a move. Part of the credit for this should obviously go to its source; the *Historia de preliis* tradition, dedicated as it is to romancing the history of Alexander, never indulges in what would represent for it a considerable detour. But we should also attend to the principle of selection that leads the poet to extract from the *Historia* just that portion of the tale that permits extended attention to just these topics. By any conventional reading, of course, the poem is incomplete— indeed, merely an extract; it begins *in medias res* ("Whanne þis weith at his wil weduring hadde, / Ful raþe rommede he rydinge þedirre," ll. 1–2) and ends with an echo of this opening, sending the "romme-riden Alixandre" off in an indeterminate direction: "þe grime king rydus. . ." (1138). At the same time, though, there are clear thematic concerns governing the extraction— that is, some store set by this particular representation of Alexander as seigneurial spokesman. We know this because *Alexander and Dindimus* essen-tially tells its truncated tale twice, supplying us with not one but two encounters between Alexander and a group of naked Indian sages.

The poem begins with Alexander's arrival at "Oxidrace. . ./ þere wilde contré was wist and wondurful peple"; the "wonderful people" are the Gymnosophists, whose exchange with Alexander, as I described at the

beginning of this chapter, ends with the latter's assertion that his Greeks are
God's "servauntus hende" and his condemnation of the Indians' egalitarian
way of life:

> ȝif God sente every gome þat goþ upon molde
> Wordliche wisdam and wittus iliche,
> Betur miȝhte no burn be þan an oþur;
> Apere miȝht þe pore to parte wiþ þe riche.
> Þanne ferde þe worlde as a feld þat ful were of bestes,
> Whan everi lud liche wel lyvede upon erþe.
>
> (101–06)

Alexander's words recall, at some distance, the seigneurial rhetoric of
Winner and Waster: "Woldest þu have lordis to lyfe as laddes on fote?. . .Late
lordes lyfe als þam liste, laddes as þam falles."[68]

After *Piers Plowman*, of course, the terms of contention can never again
be simply "lords" and "lads": Langland's *gravitas* on the topic of the "louable
lif" pulls the alliterative tradition into his orbit, and later texts, particularly
those explicitly concerned with Christian aristocratic behavior, routinely
depict the chivalric life as uncomfortably poised between the call of
the worldly and the demands of the penitential. As Hanna observes, "exem-
plarism in alliterative poems—especially in historical accounts—is always
problematic."[69] It is less problematic here than elsewhere, however;
Alexander and Dindimus, largely because of its peculiar generic heritage,
places these words in Alexander's mouth and does not then offset them with
some concluding clerical moralization or *de casibus* lament, electing instead
to give full credit to the demands of aristocratic interpellation. Alexander
claims a divine commission to conquer, but he does not do so in warlike
ignorance of his mortal limitations, as the Gymnosophists' pointed question
might lead us to expect: one could almost imagine a set of circumstances in
which his speech would be downright poignant, a dark meditation on an
aristocratic restlessness over which the worldly prince has no control:". . .I mai
stinte no stounde stille in o place / Þat I ne am temted ful tid to turn me
þennus"(97–98). In any case, Alexander's reply here shows him to be making virtue of history's necessity, and the poem, despite its simultaneous
reliance on the ascetic commonplaces that will later motivate, say, the
Alliterative *Morte*'s repentant pilgrim Sir Cradok, lets him get away with it.

   Alexander and Dindimus, then, is a "thought experiment," and one of its
cousins is Chaucer's *Knight's Tale*, another translation that takes advantage of
its pagan setting in order to genuinely interrogate the limits of the aristocratic exercise of dominion.[70] Chaucer invokes the *roman antiquite* tradition
to justify his investigation of a contemporary topic in Athenian guise, and

conducts his analysis largely through narrative; *Alexander and Dindimus*, while it too seeks to be absorbed into a romance tradition, calls on the discourse of the righteous heathen, proceeds forensically rather than narratively, and if anything is even broader in its interest (though literarily much less ambitious) than Chaucer's text: its topic is the legitimacy of two forms of life, the worldly/chivalric and the ascetic/monastic.

But "monastic" is a misleading word here. The Brahmans certainly practice the kind of self-denial that would make a Carthusian cringe, and they are cloistered in a manner of speaking, by virtue of their isolation; as Alexander observes, "For so, seggus, ȝe ben byset in an yle, / Þat þer may comen in ȝour kiþ non unkouþe peple; / Ne ȝe ne mowe of þat march in no manere wende, / But, be you loþ oþur lef, lenge þerinne" (1088–91). Nevertheless, the temptation to allegorize the poem as Skeat and others have as a shadow play between the Active Life and the Contemplative Life ought to be resisted. "Active" and "Contemplative" (and their synthesis, "Mixed") lives are after all terms from a specifically monastic discourse, a closed debate within the monastic community, rather than the poles of contrast between secular activity on the one hand and pious ascesis on the other.[71] In fact, as I have suggested, the poem has more in common with *Piers Plowman*, which counterposes in similar if much more complex ways ascetic withdrawal and worldly activity per se—uncloistered productive work. "Active" in *Alexander and Dindimus* is much closer to Langland's Haukyn, *Activa Vita*, than to monastic works of charity. And in making Alexander the spokesman for worldliness, the poem's diction once again evokes *Piers Plowman*, in six original lines inserted into his first reply to Dindimus. The Brahmans, he argues (following the Latin), are like beasts in their unreasonable refusal to take advantage of the world's opportunities. His Greeks are different:

> But we faiþful folk, þat faren as wise,
> Ben ydemed to do dedus of rihte.
> Forþy us kenneþ our kinde to acorde in trowþe,
> In swiche lawus to live þat longen to Gode,
> For to sowe and to sette in þe sadde erthe
> And oþur wordliche werk wisly to founde.
> (908–13)

The Greeks, Alexander claims, conform to the demands of "trowþe," that word of great resonance in *Piers Plowman*; here it is used to denote the norms of behavioral rectitude that conform to God's law, that is, the kind of truth that God "allows." It is one of only two such usages in the poem.[72] What is most striking about this passage, though, is its synechdochal use of plowing for "worldliche werk" in general. Plowing and not plowing come up elsewhere in *Alexander and Dindimus*; the Brahmans announce that they do not cultivate the land (293–96), and Alexander punningly replies that if they do not "erien," it must be because they have no "iren" with which to fashion plows (846–53). But in each case the poet is simply following the Latin, and in each instance plowing is just one item on a long list of literal activities that the Brahmans do not pursue. In contrast, Alexander's aside on the Greeks' "dedus of rihte" makes plowing stand in metaphorically for all the worldly activity the Brahmans reject.[73]

But Alexander is no pacific plowman; the "oþur worldliche werk" that he engages in is the work of war, the grimmest work of all. Of course, the poem has no abiding interest in moralizing that side of Alexander's reputation, despite the claims of his various interlocutors, and insofar as it tells a story, it narrates two episodes in which Alexander pointedly refrains from attacking potential (and defenseless) conquests. That, again, is the whole point of the poet's enhancement of the virtuous pagan motif—to associate secular activity with the pursuit of "trowþe" rather than the pursuit of conquest or luxury. In the world depicted by *Alexander and Dindimus* the preeminence of an aristocratic class is necessary—"God ches oþur chef kingus, / Þat scholde maistres be maad ovur mene peple," Alexander explains—but aristocratic activity in the world need not be, by its very nature, morally compromising. Worldly folk may do "dedus of rihte"—this is the argument that subtends virtuous pagan literature generally, and *Alexander and Dindimus* in particular.

The cross-fertilization of speculation about the righteous heathen and the voluminous matter of Alexander should probably be seen as inevitable; the medieval Alexander is always an exemplary figure (although he can, of course, be used to exemplify completely contrary positions), and the literature of the virtuous pagan is an exemplary discourse (though as I have demonstrated, it too varies in what it sets out to illustrate). What is striking about the conjunction, though, is the way in which the latter tradition can so easily accommodate the aristocratic preoccupations of the former. If the accounts of Alexander that have absorbed the language and motifs of the virtuous pagan scene have acquired a moral seriousness beyond the usual upper-class appeal of chivalric romances, such a gain testifies to the flexibility of the discourse of the righteous heathen. Conversely, if the Alexander texts can be seen as participating in that discourse despite the absence of spectacular salvific interventions—if Alexander is a Trajan who merely lacks a Gregory—then they begin to indicate for us the true scope of vernacular interest in the topos. The way that interest manifests itself in the specifically aristocratic genres of later medieval England is the subject of the next chapter.

CHAPTER 4

THE RHETORIC OF THE RIGHTEOUS HEATHEN

Lord Cobham's Cliché

The first parliament of Henry IV's reign, in October and November 1399, sought to consolidate the new king's power by overturning the acts of Richard's revenge parliament of 1397, restoring the lands and titles of those who had survived their condemnation and punishing those who had aided the former king in his "tyranny." Among this latter group were those who acted as appellants in 1397, bringing charges of treason against Richard's enemies, and chief targets among the appellants were three of Richard's newly created dukes, Aumale (Edward of York), Surrey and Norfolk (Richard's half-brothers, John and Thomas Holand). When the question of the dukes' fate arose among the lords on Friday, October 17—should they be arrested, as the Commons had petitioned?—the first to speak was Lord Cobham, himself only recently released from the isle of Jersey, where he had been exiled for life by Richard in 1397. Cobham's speech, delivered in what one historian calls "an atmosphere of near-hysteria generated by the accusations and counter-accusations which the bitterly divided nobles hurled at each other,"[1] adapted itself to the heightened rhetorical circumstances through a forceful contrast between the corrupt English and virtuous (if anonymous) pagans.

> The first to reply was Lord Cobham who, after speaking at some length about the evils of recent years, said, among other things, that with such a king, such leaders, and such rulers, the condition of the English people had sunk lower even than that of heathen peoples, who, although infidels to the Christian faith and thus erroneous in their beliefs, nevertheless speak the truth, acknowledge the truth, and act according to the truth. The English, however, although they are Christians and should therefore profess the truth and act

accordingly, nevertheless, for fear of the loss of their worldly goods, or of being ruined through exile, or even of death—which may befall even the constant—never dare to speak or act according to the truth under such rulers.[2]

Cobham's straightforward and uncomplicated attribution to "heathen peoples" of virtuous, righteous behavior is noteworthy, but equally interesting is what immediately follows this assertion, the concomitant separation of religious truth—that is, Christian revelation—from truthful behavior in secular matters. Though Cobham suggests that there ought to be a determinative relationship between the two—Christians, by virtue of their faith, *should* act according to truth—there clearly isn't such a correspondence among the contemporary English, while the position that their paganism should render the heathens unlikely to act virtuously doesn't seem to occur to him. The pagan kings, leaders, and rulers whose governance one might logically assume is responsible for the probity of the heathen folk do not enter into Cobham's equation at all.

One reason for these omissions is the fact that there aren't any real heathen peoples being referred to here. Cobham's heathens are a kind of moral fiction, called into being precisely by the depravity of the English, which his offhand reference throws into higher relief than a simple rebuke or denunciation could do; presumably this is why Walsingham singles out this remark "among other things." Cobham is not interested in the fate of these "infideles. . .ad fidem Christianum"; they are erroneous, as far as his remark is concerned, and likely to stay that way. Indeed, if they converted or were saved by some exceptional means they would cease to be useful to him rhetorically. Moreover, though they seem to exist in the present tense— they speak, acknowledge, and act according to the truth right now, even as the contemporary English do not—they have no real historical specificity; like the changeless Brahmans (who may in fact be lurking in the background here), they are always—indeed, always already—available to serve their exemplary function. Cobham's aside is not, in other words, an authentically ethnographic observation; it's simply a colorful cliché, the commonplace of the righteous heathen. We have already seen one version of this formula in Archpresbyter Leo's preface to the *Historia de preliis*, and another in Sir John's conversation with the Sultan in *Mandeville's Travels*; indeed, the idea that corrupt Christians compare unfavorably to morally upright pagans has a history as long as Christianity itself, and the cliché appears in many different genres and modes of writing from late antiquity through the later Middle Ages.[3]

In this chapter I want to take up some late-medieval English uses of that commonplace, though not in the interest of creating a taxonomy of its appearances, nor in order to expose it as a debased, proverbial version of the serious soteriological speculation evident in *Piers Plowman* or *St. Erkenwald*. Rather, I want to consider its aesthetic (or better, architectonic) potential. For though "conventional moralist" is about the most dismissive thing a modern critic can say about a medieval writer, it is also true that commonplaces and clichés can serve important formal and rhetorical functions, providing a structure upon which more complex arguments can be built. This is the case in fact with Cobham's reference to righteous heathens, which is a part of a larger (and thoroughly conventional) claim about the effects of bad government, a claim very much in line with the particular goals of the 1399 parliament: the vilification of Richard's rule, the justification of Henry's usurpation, and the necessity of punishment.[4] In this period we often find the rhetoric of the righteous heathen operating well outside specifically theological or even pastoral contexts, and the use of its conventional devices or motifs to effect transitions or structure episodes in a secular narrative represents, as much as does an account of an exceptional salvation, the breadth of the contemporary discourse of the virtuous pagan.

There is a certain logic to this overlap; even stories with an explicitly theological interest in virtuous pagans are typically not stories of virtuous peasants, and when a particular righteous heathen is not a dead white noble European male he is often, like Dindimus, part of an aristocracy of asceticism. These instances of borrowing between the more clerical treatments of the theme and the chief seigneurial genres of historical romance and *Fürstenspiegel* should not surprise us. What is interesting about such instances, however, is the degree to which they make obvious the dynamic as opposed to the exemplary function of the righteous heathen, the degree to which their inclusion can be seen to serve dramatic rather than doctrinal or expository goals. Certainly the Alexander texts discussed in the previous chapter offer some examples of how this formal potential was exploited in the period, and Alexander will feature in this chapter as well, specifically his appearances in Gower's *Confessio amantis*. Analyzing Gower's use of Alexander, along with the apotheosis of Chaucer's Troilus and the Gawain–Priamus episode in the Alliterative *Morte Arthure*, will let us see not only how the discourse of the virtuous pagan could be used to comment on contemporary aristocratic practices—chivalric, amatory, and political—but also how the virtuous pagan scene could help structure such commentary by providing the formal devices (the exceptional salvation, the scene of conversion, the dialogue about "lawe" or belief) that organized it.

"Swich fyn hath, lo, this Troilus for love. . ."

In considering the role of the righteous heathen in Chaucer's work I want to avoid some of the questions that animate Minnis's " 'semipelagian' lobby" and suggest instead the ways in which Chaucer shows himself to be alert to the literary possibilities—the aesthetic and organizational ones—in the exceptional salvation scenario. Minnis is doubtless correct that Chaucer does not use a particularly specialized Nominalist vocabulary,[5] but he does I think understand the grammar of the issue quite well, and that understanding is most evidently on display in the epilogue to *Troilus and Criseyde*, where what looks for all intents and purposes like an exceptional salvation—Troilus's flight to the eighth sphere—serves as a surprising structural keystone in the poem's arcing progress from Trojan past to British present.

Chaucer accomplishes this surprise in part by switching sources on us. The three stanzas devoted to Troilus's flight are adapted from Boccaccio's *Teseida*, where they describe the ascent of Arcita's soul after death.[6] This is a transposition that even the most astute readers of *Il Filostrato* (had there actually been any in Chaucer's audience) could not have anticipated, and although Troilus's death does seem partly to answer the prayer (uttered by both the narrator and Troilus himself) that disappointed lovers be permitted to pass out of this world, nothing in these prayers remotely suggests an ascent accompanied by the music of the spheres or the personal attention of Mercury as a sort of celestial concierge. It is meant to startle us in its unexpectedness and frustrate us in its vagueness—the vagueness of Troilus's final resting place, of the mechanism of his ascent, and even of his qualifications for such treatment. In the *Teseida*, Boccaccio's Arcite is not at all reticent about the reasons he thinks should earn him Elysium, but Chaucer does not import this material when he adapts the scene to his poem.[7] His failure to specify exactly the merits that qualify Troilus for his unusual treatment has helped to obscure the relevance to this scene of the contemporary debate about the salvation of the righteous heathen. Likewise the rhetorically overwrought rejection of "payens corsed olde rites" in the ensuing stanzas quickly attempts to distract us from remembering that the "fyn and guerdon" of Troilus's travail is actually a pretty good bargain: some sort of celestial residence and exactly that species of world-despising enlightenment that evidently ought to be the object of the poem's Christian readers. This orthodox, Augustinian rejection of paganism and pagan poetry thus seeks to balance the heterodox gesture—both literary and theological—that makes Troilus into a saved pagan soul.[8]

That sounds like an overstatement, though in making it I consider myself in good company.[9] After all, Chaucer is determinedly vague about just where "Mercurye sorted him to dwelle." Moreover, criticism of *Troilus and Criseyde* has generally downplayed the theological ramifications of Troilus's apotheosis, perhaps in part because the epilogue causes enough trouble already for a non-exegetical reading of the poem. As Winthrop Wetherbee writes, "we do not need to engage in speculation about the salvation of the righteous heathen to find value in an experience of poetry which withholds the final orientation of a religious perspective until we have been made to see in the condition of Troilus. . . .the lineaments of the love that upgroweth in Chaucer's hearers."[10] But whether we need the righteous heathen or not, that is exactly what we get in the 1380s, a decade in which *Piers Plowman* and *Mandeville's Travels*, both of which Chaucer certainly knew, were circulating their own accounts of virtuous pagans, and in which the mid-century controversialist writings of Thomas Bradwardine (to whom Chaucer refers in the *Nun's Priest's Tale*), Robert Holcot (whose Wisdom commentary was a source for that tale), Uthred de Boldon, and others—including Wyclif— had clearly entered vernacular debate, and Chaucer's purview.[11] Indeed, a few years after *Troilus and Criseyde*, *St. Erkenwald* would specifically link the topic of the righteous heathen with Trojan or neo-Trojan history by identifying its virtuous pagan judge as a citizen of Troynovaunt, the city allegedly founded by Aeneas's grandson Brutus and later known as London. *Erkenwald*, as we have seen, is often considered a theologically conservative response to the liberal view of exceptional salvation found in *Piers Plowman*—but not even Langland is as bold as Chaucer, launching Troilus skyward without even a polite nod to his virtue or justice or faithfulness to his pagan law. In that sense, Troilus is not one of the usual suspects.

Or maybe he is. He is after all a convert, having gone from a scorner of Love's "lewed observaunces" (1.198) to the god's most ardent, if initially most hapless, servant; as Pandarus observes,

I thenke, that sith Love, of his goodnesse,
Hath the converted out of wikkednesse,
That thou shalt ben the beste post, I leve,
Of al his lay, and moost his foos to greve.
(1.998–1001)

The implicit allusion here is to the conversion of Saul, and the larger frame of reference is of course the "Religion of Love" so characteristic of courtly romance from the twelfth century onward, tricked out in *Troilus* as elsewhere in the tradition with the language of the sacraments and of grace (e.g., Troilus's "confessions" to the God of Love at 1.932–43 and 2.522–43). Windeatt succinctly describes the motif: "The relationship between lover and lady, the suppliant and the source of all mercy, is structured by frequent analogy with the Christian concept of grace, while the eventual attainment of sexual fulfillment is set about with references to heaven. The familiar structures and aims of religious devotion in this way act as a correlative to the devotion within the love-affair, lending it the force and value of religious associations, yet also cumulatively suggesting the ways in which Troilus' love is not in itself a religion."[12] But this correlative approach to religious form and devotion, as we have seen in the case of *Mandeville's Travels*, can lead to some very liberal-minded conclusions about the fate of pagan folk; devotion per se can sometimes compensate for a certain degree of theological waywardness, which is what seems to be happening in the case of Troilus's ascent. That is, from the perspective of the end of the poem, Troilus's conversion and his continued devotion to the "lay" of the God of Love can be taken as episodes in the biography of a righteous heathen; the ascent to the heavens retrospectively recruits these moments from one conventional model to another, from the allegorical religion of love to the discourse of the virtuous pagan. It is as if a passage from the *Romance of the Rose* has been interrupted by an excerpt from *Piers Plowman*—and of course we know from *Piers* that the willingness to convert to a better law is a distinguishing mark of the righteous heathen ("And if þer were he wolde amende, and in swich wille deieþ"). What's relevant in this analogy, it should be stressed, is the structure of Troilus's devotional life, not its content: that structure would have borne, to one of Chaucer's contemporaries, a strong family resemblance to a tale of a righteous heathen, but contemporary readers would also have recognized Troilus's pagan limitations, as Minnis and many others have argued. Troilus's Boethian and Dantean hymns to the binding power of love in book 3 do not ultimately qualify him as enlightened, since they are followed by the plaintive philosophical fatalism of books 4 and 5. Troilus is no Dindimus.[13] But thinking about the poem's conclusion in terms of structure makes clear the degree to which it is asking us to think in terms of analogy: analogies between Troilus and other righteous heathens, between *Troilus and Criseyde* and the "poesye" of Chaucer's classical masters, and between "the condition of Troilus. . .[and] the lineaments of the love that upgroweth in Chaucer's hearers."

Indeed, the question of Troilus's ascent and the mystery of his ultimate destination has typically been discussed in terms of analogous episodes elsewhere, whether in Dante or Lucan or the Boethian flight of the mind or the transmogrified civic piety of Statius's Menoeceus. To this list we can confidently add one more literary analogue that probably would have occurred to that large majority of his early readers and auditors untutored in Dante or ignorant of Boccaccio, but passably familiar with the work of Chaucer himself:

> Thanne telleth it that, from a sterry place,
> How Affrycan hath him Cartage shewed,
> And warnede him beforn of al his grace,
> And seyde him what man, lered other lewed,
> That lovede commune profyt, wel ithewed,
> He shulde into a blysful place wende
> There as joye is that last withouten ende.

Thanne axede he if folk that here been dede
Han lyf and dwellynge in another place.
And Affrycan seyde, "Ye, withouten drede,"
And that oure present worldes lyves space
Nis but a maner deth, what wey we trace;
And rightful folk shul gon, after they dye,
To hevene; and shewede hym the Galaxye.

Thanne shewede he hym the lytel erthe that here is,
At regard of hevenes quantite;
And after shewede he hym the nyne speres;
And after that the melodye herde he
That cometh of thilke speres thryes thre,
That welle is of musick and melodye
In this world here, and cause of armonye.

Than bad he hym, syn erthe was so lyte,
And dissevable and ful of harde grace,
The he ne shulde hym in the world delyte. . . .
(PF 43–66)

All the *topoi* of Troilus's heavenly ascent—the guided journey; the little
earth; the music of the spheres; the *contemptus mundi* perspective—are here
unambiguously previewed in the *Parliament of Fowls*, in the same rhyme royal
stanza voiced by the same loveless narrator as in *Troilus and Criseyde*,
engaged in a similar kind of translation project, recounting not Lollius but
Macrobius.[14] Indeed, the passage even contains a proleptic answer for readers
who might prefer to view Troilus's love as a conventionally damning idolatry
rather than potentially salvific devotion:

But brekers of the lawe, soth to seyne,
And likerous folk, after that they ben dede,
Shul whirl about th'erthe alwey in peyne,
Tyl many a world be passed, out of drede,
And than, foryeven al hir wikked dede,
Than shul they come into that blysful place,
To which to comen God the sende his grace.
(PF 78–84)

Without attributing to Chaucer an ongoing belief in postmortem revelation or a Uthredian *clara visio*, we can nevertheless say that in the 1380s,
both in his poetry and in the literary environment in which it circulated,
the *topos* of the virtuous pagan was both relevant and active, and that in
his poetry specifically, an ascent to the heavens was generally a good thing.

Troilus's ultimate destination may be mysterious, his ascent ambiguous—but given the context it's not *that* ambiguous.[15]

But having established an enlarged contemporary context for Troilus's ascent, we still haven't explained its function in the poem, beyond a somewhat fuzzy feeling that Chaucer wished Troilus well. Here again it can be useful to think in terms of structure, of the placement of Troilus's ascent not just at the end of his life (where logically it must go) but at the end of Chaucer's poem, where its status as the last event of the story is entirely contingent: the story doesn't have to end here, as Henryson, for example, would later make appallingly clear. The event itself, that is, is not in any logical sense the whole point of the poem: however closely Troilus's ascent resembles an exceptional salvation, it is not ultimately the result of explicit divine intervention, and Troilus, unlike Trajan, was not the subject of six centuries' worth of clerical and vernacular speculation. It is not *Deus artifex* but the *artifex* himself who makes Troilus fly where and when he does. Why?

One does not have to subscribe fully to Donaldson's characterization of the end of the poem as "a kind of nervous breakdown in poetry"[16] to see the last hundred lines as preserving a sequence of provisional and patently unsuccessful attempts at bringing things to a conclusion. As Patterson observes, "Criticism has often described what Lowes long ago called 'the tumultuous hitherings and thitherings of mood and matter in the last dozen stanzas of the poem,' descriptions that provide vivid evidence of the interpretive impasse to which the poem brings both its narrator and its readers."[17] That impasse arises, to put it most simply, from a problem of transition: how does a poet so deeply engaged with his portrait of pagan Troy that generations of critics have been induced to identify the narrator as virtually a fourth major character manage to disengage himself from that world and return to his own present in a way that won't do a kind of rhetorical violence to the preceding 8000 lines?[18] And the solution to that problem is Troilus's apotheosis, which is what formally enables this transition to take place. If, as I have argued, literary representations of the righteous heathen are more than just exercises in popular theological casuistry—if they express an urgent cultural need to explore and understand the possibility of an ethical genealogy that connects the pre-Christian past with the post-revelation present—then thematically Chaucer's description of Troilus's mysterious flight occurs just at the point we might expect it to, at the moment when the poem moves definitively forward from its dream of pagan Troy to fourteenth-century England. And if, as we have seen with *Piers Plowman* and *St. Erkenwald*, the representation of an exceptional salvation can serve to break a narrative impasse, then structurally Troilus's apotheosis comes at the right moment too. Indeed, its appearance begins to seem overdetermined.

After first trying, with calculated humility, to represent his "makyng" as an episode in the history of rhetoric—placing it, that is, between the "poesye" of "Virgile, Ovide, Omer, Lucan, and [emphatically] Stace" in the past, and the potentially mismetering and misreading practices of the future—Chaucer turns from the place of the book-as-object to the place of the book's subject in the history of cultures.

> The wrath, as I bigan yow for to seye,
> Of Troilus the Grekis boughten deere. . .
> But—weilawey, save only Goddes wille,
> Despitously hym slough the fierse Achille.
> (5.1800–01, 1805–06)

The Homeric allusions in this stanza, which begins with wrath and ends with Achilles, locate Troilus at the moment of his death firmly in a classical setting, literally and rhetorically. Over the next several stanzas, however, as Troilus moves from the battlefield to the eighth sphere and beyond, we move with him from the past to the Christian present of the poem's composition and reception. The rhetorical gambits alone are not in themselves sufficient to make the transition possible; the gesture of respect toward classical predecessors—poetic *pietas*—has to be joined to a gesture of recuperation as well; once again, *pietas* and piety are superimposed, as in Gregory's act on Trajan's behalf. Thus the same gesture that individually recuperates Troilus for an anachronistic Neoplatonism also (and analogously) prepares the poem for consumption by medieval Christian readers, and for the corrections of Gower and Strode, and for the concluding prayer to "that sothfast Crist." To put it another way, the *translatio* of Troilus, when seen in the context of contemporary speculation about the righteous heathen, accomplishes in small what the "epilogue" of *Troilus and Criseyde* does for the poem as a whole, and its subject: it both recuperates (in the sense of making ideologically intelligible) and recovers (in the sense of claiming as one's own, out of desire and esteem) the Trojan past, its heroes and ladies and its anachronistically courtly and romantic culture.[19]

Sir Gawain and the Greek Knight

A similar gesture of recovery helps to structure the Alliterative *Morte Arthure*. Geraldine Heng has described the *Morte's* "omnivorous appetite for bodies of disparate matter—heraldry, law, diplomacy, government, cookery, nautical warfare, a spectrum of linguistic imports and innovations"—and the rich stew the poem thus represents for critics.[20] Like several of its alliterative cousins, the poem also partakes of the discourse of the virtuous pagan, though unlike *Piers* or *St. Erkenwald* the poem eschews sacramental or soteriological questions in favor of a more straightforwardly secular and chivalric use of the righteous heathen motif. Indeed, though the poem begins with a conventional prayer that "glorious Godde. . .gyffe vs grace to gye and gouerne vs here / In this wrechyde werlde thorowe vert[u]ous lywynge, / That we may kayre til Hys courte, the kyngdom of hevyne, / When oure saules schall parte and sundyre fra the body," the *Morte* is notably reluctant to represent instances of such grace, despite the large number of souls sundered from bodies in the course of its narrative.[21] Even the final scene of Arthur's requiem mass exhibits more interest in the mourners' elegant funeral attire than in speculations about the fate of the king's soul. But, if the poem shows scant interest in conventional salvations, much less exceptional ones, it is quite concerned to explore the more broadly recuperative gestures that the discourse of the righteous heathen underwrites, and the *Morte*-author's adaptation of the virtuous pagan scene is one of the period's most inventive explorations of its formal and thematic possibilities.

The scene in question is the encounter between Sir Gawain and Sir Priamus, an episode the *Morte*-author adds to his largely Galfredian narrative. Their joust and subsequent dialogue are in fact already inserted into a larger addition that frames their interaction: just past the halfway point of the poem, in the course of Arthur's apparently inexorable progress toward Rome, Gawain and others are sent off on a foraging expedition (based on the French Alexander romance *Li Fuerres de Gadres*), during which Gawain meets a lone knight whom he inevitably challenges to battle (an episode drawn from the Middle English Charlemagne romance *Sir Ferumbras*).[22] If the marvelously bloody single combat between Gawain and Priamus takes us into a realm of pure romance—as Chism and others have observed, the scene "is almost parodically crammed with romance conventions"[23]—the conversation that ensues swerves into virtuous pagan territory. In furious and lavishly detailed battle, Gawain lays his adversary open to the liver, while he himself is nearly bled white; they pause, and then Gawain's opponent opens negotiations. I promise to heal you, the unknown knight says to Gawain, on the condition that "thow suffre me, for sake of thy Cryste, / To schewe schortly my schrifte and schape for myn end" (2587–86). These words are not entirely unambiguous—petitioning by "thy Criste," suggests his alienation from the Christian faith, at the same time that he seems to ask for its conventionally sacramental last rites—but Gawain's reply leaves no doubt about his own assumptions. Like Erkenwald before the tomb and Alexander writing to Dindimus, he asks the knight to identify his "laye," the spiritual law that he follows:

> I gyfe the grace and graunt, þofe þou hafe grefe seruede—
> With-thy thowe say me sothe what thowe here sekes,

Thus sengilly and sulayne all thi selfe one;
And whate laye thou leues one, layne noghte þe sothe,
And [vndir] whate legyaunce and whare þow arte lorde.
(*MA* 2590–94)

The mystery knight reveals himself to be Sir Priamus, the son of a prince, a descendant of four of the Nine Worthies (Alexander, Hector, Judas Maccabeus, and Joshua), and heir to Alexandria and Africa—a fantastical genealogy that identifies him, as Lee Patterson has persuasively argued, as "a classical warrior, heir to the heroic virtues of an antique world."[24]

The addition of this heritage is one of two important alterations the *Morte*-author makes to his source. In *Sir Ferumbras*, the eponymous antagonist of Charlemagne's vassal Oliver is not a faux-Trojan princeling but a stock Saracen giant, son of Sultan Balan. In the aftermath of that battle, Ferumbras is converted to the one true faith so wholeheartedly that he later accedes to the execution of his own unregenerate father. But in the *Morte*, the conversion is omitted—or rather, it is rendered nonsacramentally: Priamus and his retinue immediately join Gawain's forces (providing valuable intelligence about an imminent ambush by his erstwhile ally the Duke of Lorraine), enacting what Patterson calls "a brief allegory of the transactions of past and present" that depicts "the appropriation of the old, Alexandrian values by the new order."[25] Priamus himself, like Mandeville's sultan, acknowledges that he is living at the moment of supercession, that with such soldiers as Gawain, Arthur "will be Alexander ayre, that all þe erthe lowtedde, / Abillere þan euere was sir Ector of Troye!" (2634–35). This relatively uncomplicated act of *translatio* has, in Patterson's reading of the *Morte*, relatively unhappy consequences; the poem's unwillingness to sacralize Priamus's virtues here, or the virtues of the Worthies who appear in Arthur's dream of Fortune, suggests to him that "Classical heroism is transferred but not transformed, and no answer is given to the question of how Arthur, invested with Alexander's force, can avoid Alexander's fate." Christianity, so often the bright line in virtuous pagan stories, turns out to be a distinction without a difference here.[26]

Given such a dour conclusion—one largely justified by the events of the poem, I think—we might be tempted to ask why this scene should be included under the virtuous pagan rubric at all. But in ca. 1400, the generally accepted date of the *Morte*, any scene that turns on the interrogation of an unknown "laye" necessarily places us in the realm of the righteous heathen.[27] The motif as it is employed here is structurally analogous to the *Erkenwald* model, devoid of any immediate sacramental content and focused instead on martial virtues—Trajan's story rendered as a *bel inconnu* romance. Like Troilus, Priamus is no Dindimus, since it's clear that the "laye" that he follows is the law of chivalry, but his thoroughgoing devotion to it—amply manifested in his prowess as well as his heritage—is sufficient to vouchsafe him the kind of cultural approbation represented by his absorption into Arthur's army.

It turns out, though, that the Priamus scene has more structural analogues than just *St. Erkenwald*, and that the salvific concerns characteristic of a virtuous pagan incident are not really absent from the poem. They've been displaced onto a later scene—Arthur's encounter with the pilgrim-knight Sir Cradoke—which is one of three striking moments in the *Morte* that self-consciously reflect not only on the romance mode in which they are represented but on the self-contradictions of the chivalric ideology they exemplify, giving a structural and thematic integrity to the poem's ostensibly historical narrative. The three episodes are Arthur's battle with the giant of Mont St. Michel, in which the destructive violence of the king's imperialist ambitions are reflected back to him in savage form; Gawain's encounter with Priamus, in which the Christian chivalry of the former reveals its essentially untroubled kinship with the pagan chivalry of the latter; and Arthur's meeting with Cradoke in the aftermath of his dream of Fortune, in which the philosopher's exposition of that dream's imperative—"Schryfe the of thy schame and schape for thyn ende!" (3400)—is rendered as narrative in the encounter between the knight-as-worldly-conqueror and the knight-as-penitential-pilgrim. The first and the last of these scenes have often been paired in the criticism, as they both involve Arthur, but Arthur haunts the middle scene, too: Gawain may become Priamus's friend and companion, but Arthur will be his heir, as the latest of the Worthies. A number of formal details and devices are repeated in these scenes, connecting them to one another. All are animated, at least at first, by a certain obstreperousness: the brutish giant seeks Arthur's life, Priamus proffers battle, and Cradoke is aggressively unimpressed by Arthur's warning that he is passing through a war zone: "I will not wonde for no werre to wende whare me likes" (3494). In addition, all three episodes contain instances of concealed identity or misrecognition: the "woful widow" mourning her foster-daughter, who has become the giant's most recent victim, does not recognize Arthur, who at first claims to be one of his own knights; Gawain initially hides his identity from Priamus, pretending to be a knave-made-yeoman; Craddock fails to recognize Arthur at all, and vice versa. Likewise they are all preceded by "locus amoenus" garden scenes that signal their irrelevance to the narrative's Roman plot (a motif that, in the third episode, is displaced into the dream of Fortune). The first two episodes feature extended single combat largely unrelated to the progress of the war, while the first and the third immediately follow prophetic dreams, attend specifically to Arthur's arming or dressing himself, and employ the language of pilgrimage (ironically, in the Mont St. Michel scene).[28]

In each of these solitary encounters, then, a chivalric Christian knight meets his double and is given a chance—in theory, anyway—to step outside the narrative and recognize his ideology's various contradictions. As a kind of structural refrain in the poem, these moments tend to corroborate critical readings that consider the *Morte* an extended meditation on and critique of aristocratic practices, a virtual transcription of late-medieval chivalry's attempt at the talking cure, rather than as a straightforward tragedy of Fortune or a sanguinary indictment of *surquidrie*.[29] And it is the addition of the Priamus episode—its construction on the chassis of the *Ferumbras* story and its inclusion toward the middle of the *Morte*—that serves as the lynchpin of such readings, for it is the only one of these three scenes imported into the poem from elsewhere; both of the others are ultimately inherited from the *Brut*-tradition on which the poem is based, though they appear in the *Morte* in modified forms (forms that align them more fully with the Priamus episode, in fact).[30] It is the middle scene that directs us to understand the other two relationally, that invites us to reread the battle with the giant as more than just a marvelous digression and the meeting with Cradoke as something other than the necessary arrival of a *nuntius*, that induces us to group these three scenes together as representing Christian chivalry's encounter with its savage prehistory, its classical past, and its contemporary spiritual imperatives. Importing the combat/conversion episode from the *Ferumbras* tradition into the *Morte* is therefore doubly analogous to what Chaucer does at the end of *Troilus*. Not only do both writers supplement one historical romance with an excerpt from another—an analogy obscured, perhaps, by the fact that both of Chaucer's sources just happen to be by Boccaccio—but each also supplements by borrowing elements of the virtuous pagan scene, one more example of the structural utility of that scenario as an aesthetic resource.

Nominalist Gower?

Gower takes an entirely conventional approach to the topic of conversion in the *Confessio amantis*; his theology is essentially conservative as regards salvation, and he seems not particularly interested in exceptional circumstances. At the end of his survey of world religions in Book 5 of the *Confessio*—to Macaulay, famously, a "very ill-advised digression"[31]—he makes quite clear the stipulations for salvation, a perfectly unexceptionable combination of faith (the correct faith) and works:

> So stant the feith upon believe,
> Without which mai non achieve
> To gete him to Paradis ayein;

Bot this believe is so certein,
So full of grace and of vertu,
That what man clepeth to Jhesu
In clene life forthwith good dede,
He mai noght fail of hevene mede,
Which taken hath the rihte feith;
For elles, as the gospel seith,
Salvacion ther mai be non.
And forto preche therupon
Crist bad to hise Apostles alle,
The whos pouer as nou is falle
On ous that ben of holi cherche,
If we the goode dedes werche;
For feith only sufficeth noght,
Bot if good dede also be wroght.
                    (*CA* V.1785–1802)

The marginal gloss here is James 2:20, "Fides sine operibus mortua est," though it would probably be a mistake to assume that the emphasis on works betrays any semi-Pelagianism in Gower's soteriology. The account of the four predecessors of the Christian faith—the religions of the Chaldeans, Egyptians, Greeks, and Jews—concerns itself more with "misbelieve" than misdeeds.

At the same time there is also evidence in the *Confessio* that Gower was in step with contemporary ideas about the virtuous pagan figure, even if he is more modest than Langland in his representations. As Larry Scanlon has observed, in adapting the tale of Constantine and Sylvester for the end of book 2, Gower "changes Constantine from a infidel persecutor redeemed, like Paul on the road to Damascus, to a type of the virtuous pagan miraculously granted enlightenment. . .Constantine's conversion begins from within the temporal and the social, with the engendering of 'pite' within his own heart."[32] That pity arises when Constantine hears the weeping and lamentations of the mothers whose young children he has caused to be rounded up, in response to the conclusion of his advisors that the only cure for Constantine's sudden attack of leprosy is a bath "in childes blod/ Withinne sevene wynter age" (*CA* 2.3207–08). Moved, like Chaucer's Theseus or the *Legenda*'s Trajan, by the tears of women, he awakes as if out of a dream and remembers the "divine pourveance" that has equally ensouled the "povere child" and "the kinges Sone," feeling such pity that "him was levere forto chese / His oghne body forto lese, / Than se so gret a moerdre wroght / Upon the blod which gulteth noght" (2.3291–94). He releases the women and their children, compensating them with a portion of his treasury, and presumably settles in for a life of leprous isolation. God has other plans for him, however:

> Bot now hierafter thou schalt hiere
> What god hath wroght in this matiere,
> As he which doth al equite.
> To him that wroghte charite
> He was ayeinward charitous,
> And to pite he was pitous:
> For it was nevere knowe yit
> That charite goth unacquit.
>
> (2.3325–32)

"Ne wolde neuere trewe god but trewe truþe were allowed": God sends Constantine a dream that very night in which Peter and Paul advise him to seek out Sylvester, who has been in hiding for fear of Constantine's persecution of Christians; Sylvester will provide "enformacioun" pertaining to his salvation. The second half of the tale plays out parallel to the first: Constantine's men round up Sylvester, the emperor hears, instead of weeping, his catechism, and the vessel that was to have held the children's blood is used for his baptismal submersion, during which the scales fall from the body of the now enlightened emperor.[33]

There is nothing particularly exceptional about this route to salvation, which in most ways is an orthodox conversion narrative. Nevertheless, it shares enough details with the paradigmatic Trajan episode—an emperor interrupted in his plans by feminine complaint, a papal intervention, a divine message about the rewards of pity—to remind us of the *Legenda* account, which may have been alive in Gower's imagination too.[34] In fact the episode is a perfect working-out of the *facere quod in se est* principle; the pagan Constantine does what is in him (by recognizing the *divine pourveance* and acting with pity), and God does not deny him grace—though rather than salvific grace via the *potentia absoluta*, it's a prevenient sort that sends Constantine to the tutor who can most effectively bring to him the fruits of revelation and an opportunity to experience the greater grace that comes with sincere conversion to the one true faith. Gower's implicit reliance on the *facere* principle, in fact, is an index of its relative orthodoxy, and is consistent with the "medieval 'classicism' "[35] that is his perspective throughout the *Confessio*—a text after all that adapts pagan stories to a Christian frame narrative, the lover's confession offered according to the rubric of the seven deadly sins.

Having decided that spectacular salvations held little appeal for Gower, however, there still remains the structural question. That is, does Gower ever use the virtuous pagan theme to stage an important transition in his poem,

as with Troilus's apotheosis or Priamus's combat? It is tempting to put the
Constantine tale in this category; not only does it finish book 2's survey of
Envy, but it also closes off a lengthy exploration of the right relations
between lay and ecclesiastical power that begins with the preceding tale
of the corrupt Pope Boniface, who is eventually deposed by the king of
France, and concludes after Constantine's conversion with the Donation of
Constantine, an event that Genius very conventionally deplores ("Today is
venym shad/ In holi cherche of temporal / Which medleth with the spiral,"
2.3490–92).[36] But in fact there is better evidence for the claim that Gower,
like Chaucer, understood the structural possibilities of the discourse of the
righteous heathen, and although neither the context nor the content is
specifically salvific, it is from a certain point of view catechetical. I refer to
the role of book VII, the Confessio's *Fürstenspiegel*, which Gower specifically
represents as an exchange between two pagans, Alexander the Great and his
tutor Aristotle.

Gower's Alexander

Gower's portrait of Alexander in the *Confessio* has been judged inconsistent
or, alternately, self-undermining, largely due to the contrast between
his appearances in book III ("Alexander and Diogenes," 3.1201–1330;
"Alexander and the Pirate," 3.2363–2437; "Wars and Death of Alexander,"
3.2438–80) and his role as Aristotle's pupil in the seventh book. James
Simpson and Diane Watt have both argued that, since the earlier incidents
actually represent later portions of Alexander's life story, they indicate the
apparent failure of Aristotle's attempt to educate the prince, who has evidently not been properly "enformed" by his tutelage.[37] Watt is certainly
correct in observing that trying to read Alexander's several appearances as "a
narrative of progress, a medieval bildungsroman" presents a number of
problems, but it would be a mistake to conclude with Cary that the tradition of his education by Aristotle "had no apparent effect upon the current
conception of Alexander," and that "the most living side of the English late
medieval conception of Alexander is represented by the thorough condemnation of him by Gower and Lydgate."[38] In fact Gower's work manifests a
fundamental ambivalence about Alexander's exemplary role, because Gower
is himself a victim of Alexander's ubiquity in late-medieval England, that is,
a victim of his several sources, which as we have seen in the previous
chapter offer their readers several moral positions.

Unlike the author of *Mandeville's Travels*, whose Alexander comes mostly
from romance histories, Gower ranged much more widely: for the
Nectanabus story in book VI he drew primarily on the *Roman de Toute
Chevalerie* of Thomas of Kent, a late twelfth-century verse romance based

on the *Epitome* of Julius Valerius, while elsewhere in the *Confessio* he can be seen borrowing from the *Historia de Preliis* and exemplum-collections like that of Valerius Maximus.[39] Book VII, of course, draws on the tradition of the *Secretum secretorum*, allegedly a letter sent by Aristotle to Alexander (busily engaged on a distant campaign) that describes not only the rules and requirements of good kingship but also the dietary and hygienic regimen necessary for a healthy body.[40] These sources vary not only in content and moral but in mode, offering narrative, exemplary, and expository perspectives on Alexander that resist being assembled into any sort of integrated portrait. Indeed, in some ways Alexander is a perfect subject for the *Confessio* precisely because of his multifarious meanings; critics who find a productive tension between the poem's narratives and the Latin glosses meant to moralize them might find such a conflict embodied in Alexander all by himself.[41]

At the same time it is possible to see Gower taking steps to lend some consistency to his representation, reading the earlier appearances not as episodes in a failed narrative of moral progress but as preparations for the governing conceit of book 7. Several of the anecdotes stress Alexander's educability. In the Diogenes episode in particular, Gower goes out of his way to depict Alexander as patient and courteous to the old philosopher. When he sends a knight to investigate the singular spectacle of Diogenes seated in his rotating tun, the knight is made "riht wroth" by the philosopher's apparent insolence and insults him. In contrast, Alexander, "which hadde wordes wise, / His age wolde noght despise" (III.1263–64). Diogenes calls the conqueror "mi mannes man" rather than "mi king," and when he is challenged for this seeming disrespect defends himself by observing that he rules his own will through reason while Alexander's reason is subject to Alexander's will.

> The king of that he thus answered
> Was nothing wroth, bot whanne he herde
> The hihe wisdom which he seide,
> With goodly wordes thus he preide,
> That he hem wolde telle his name.
> (III. 1293–97)

When Diogenes reveals his name, Alexander is delighted to find himself in the presence of the philosopher he has already heard so much about. The two of them enact the Gymnosophist paradigm, as Alexander promises to do a boon to Diogenes, and the old man asks him to move out of his light. More sinned against than sinning in this vignette, Alexander suffers here because of the episode's structure rather than because of the content of his character: at no time, in fact, does he evince any of the behavior upon which Diogenes bases his remark. His reputation not only precedes him; it literally replaces him.[42]

Likewise, the anecdote of the Pirate (who, when brought to judgement before Alexander, claims that the difference between them is one of degree, not of kind—that Alexander is called a conqueror and not a thief only because of his great armies) shows an Alexander who is at least self-knowing; he acknowledges the "wordes wise" of the Pirate, and rewards rather than condemns him by taking him into his service. This latter anecdote arises in the course of a discussion of murder and the evils of war, which Genius concludes by relating a brief history of Alexander's career and reading the inescapable Wheel-of-Fortune moral into it. Alexander's problem is "that reson mihte him non governe," so that "as he hath the world mistimed, / Noght as he scholde with his wit, / Noght as he wolde it was aquit" (3.2443, 2458–60); at the height of his glory and achievement, and in the middle of his own empire, "he was deceived, / And with strong puison envenimed" (3.2456–57). Certainly Genius's conclusion here is harsher than that of the narrator of Chaucer's *Monk's Tale*, for whom Alexander's death cues an apostrophe against "False Fortune" (VII.2669). But it also bespeaks some of the tensions inevitable in *de casibus* accounts of Alexander that have to trade on his worldly prestige in order to draw a sufficiently affecting moral from his demise. Wetherbee's diagnosis, that "the world of chivalry is for Gower an uncentered world of ceaseless, random movement, its activities often directly at odds with the social order," seems apt here.[43] The "Alexander and the Pirate" story is symptomatically paradoxical: the narrative proper ends with the laudatory observation that the Pirate, once in Alexander's service, became "an orped kniht" who "gret prouesce of armes dede, / As the croniques it recorden," but Genius follows up this remark with an immediate rebuke: "And in this wise thei acorden, / The whiche of o condicioun / Be set upon destruccioun" (3.2415–20).

Gower shows similar evasiveness again in book 5, where he reproduces a portion of Dindimus's correspondence with Alexander about the idolatrous practices of the Greeks, describing how they have a deity for each part of the body (5.1453–96). This passage, derived from the *Historia de Preliis* tradition, mentions Alexander only once, as the recipient of Dindimus's letter; the conqueror is thus detached to a degree from the disparagement of the recital, which is largely a coda to the 610-line account of the Greek religion. (It's also a sign that Gower was fully acquainted with the Brahman tradition, though this is his only mention of it in the poem.) A few dozen lines later, in a passage on the origins of idol-worship, Genius recounts how Alexander was made to hear an image of Serapis speak "thurgh the fendes sleihte," and how

> thus the fend fro dai to dai
> The worschipe of ydolatrie
> Drowh forth upon the fantasie
> Of hem that weren thanne blinde
> And couthen noght the trouthe finde.
> (V. 1586–90)

Alexander's motive here is the pursuit of truth; when he first hears about the speaking idol, he desires to "have / a knowlechinge if it be soth" (1578–79). Blame is diluted by reference to the fiend, and Alexander's fault turns out to be the ignorance charateristic of his era (i.e., "of hem that weren thanne blinde") rather than willfulness. Finally, Alexander apparently incurs no blame for his murder of Nectanabus at the end of book 6. One would think that, had Gower been looking for an opportunity to demonstrate "that reson mihte him non governe," this would be it, but Genius reserves judgment for Nectanabus alone, declaring that "for o mis an other mis / Was yolde, and so fulofte it is; / Nectanabus his crafte miswente, / So it misfell him er he wente" (6.2359–62). Thoughout the story, Alexander is largely defined by Nectanabus's prophecies, both that he "schal winne / The world and al that is withinne" (6.2165–66) and that ultimately "he schal with puison deie" (6.2246)—prophecies whose truth has already been confirmed for us in book 3.

Thus Gower prepares us for the well-educated Alexander of book 7 by regularly softening the evaluative force of the earlier episodes, which would presumably permit him to take full advantage of the prestigious pairing of the conqueror and the philosopher. One critic claims that "No two figures of Ruler and Sage were more common as medieval archetypes than those of Alexander and Aristotle,"[44] and their relationship offered both a model and a precedent for interactions between a sagacious counsellor (or poet) and a prince ostensibly interested in virtuous behavior—an originary moment for the *Fürstenspiegel* tradition, which had the additional advantage of being based on historical fact.[45] Almost every time Genius introduces a topic in book 7, Gower has him remind Amans and the reader of the historical situation from which he is ostensibly retrieving this material: "Bot ferst, as it was forto done, / This Aristotle in other thing / Unto this worthi yonge king / The kind of every element / Which stant under the firmament, / Hou it is mad and in what wise, / Fro point to point he gan devise" (VII. 196–202; see also 710–20, 1271–80, 1295–1303 and Latin gloss, 1645–46, 1699–1703, 1726–33, 1976–79, 2031–35 and gloss, 3084–91, 4233–37 and gloss, 4257–61 and gloss, 4558–60 and gloss, 5384).[46] Thus, the validity of book 7's pedagogy is constantly being checked against its alleged exemplar.[47]

At the same time, though, in a move that has inspired considerable critical commentary, Gower also presents the turn to the *Secretum* material as entirely contingent, based on what seems like little more than Amans's whim. Genius first mentions the pedagogical relationship between the two in the context of the Nectanabus story, where he notes that in Alexander's youth "Calistre and Aristote / To techen him philosophie / Entenden. . ." (VI. 2274–76).[48] A little over a hundred lines later, Amans asks his confessor for further information:

> Bot this I wolde of you beseche,
> Beside that me stant of love,
> As I you herde speke above
> Hou Alisandre was betawht
> To Aristotle, and so wel tawht
> Of al that to a king belongeth,
> Wherof min herte sore longeth
> To wite what it wolde mene.
> For be reson I wolde wene
> That if I herde of thinges strange
> Yit for a time it scholde change
> Mi peine, and lisse me som diel.
>   (VI. 2408–19)

Critics disagree about the effectiveness of Gower's transition here; one praises "the adroit way in which he manoeuvred from love in Book VI to 'policie' in Book VII and back again to love in Book VIII," while another suggests that the use of "a long lecture on higher education as a tranquilizer against erotic anxiety teeters on the brink of the absurd. . . ."[49] Such difference of opinion may arise in part from the fact that Gower makes something of a joke in this passage, first by allowing Amans to extrapolate "al that to a king belongeth" from Genius's mere "Philosophie," and second by having him characterize as "thinges strange" the story of Aristotle's tutoring of Alexander, a relationship memorialized in one of the most popular secular books of the later Middle Ages.[50] But it is also the case that familiar, weighty matter is introduced here in a particularly oblique fashion, not with a flourish but with Genius's complaint that he is "somdel. . .destrauht" to find himself responsible for presenting matter "noght in the registre / Of Venus" (7.6, 18–19). Gower seems to highlight both the authority that book 7 derives from its putative source and the apparent unfamiliarity of that source to his poem's interlocutors.

What most interests me about the breaking of the penitential frame here is the conceit that appears to govern it, Gower's choice to represent the ideal—indeed, the originary—princely education as an encounter between two virtuous pagans. Certainly the *Secretum* tradition was widely familiar (whatever Amans claims) and accepted as authentic, but at the same time we know that Gower depended for the actual organization and contents of book 7 much more heavily on two other sources, Brunetto Latini's *Tresor* and the *De Regimine Principum* of Aegidius Romanus.[51] Both of these works are thoroughly Aristotelian and cite the philosopher frequently, but neither is presented as a transcript of an Aristotlelian text: both are addressed to contemporary Christian readers in the author's own voice, Aegidius directing his to Philip the Fair and Latini to a "friend." It's Gower who stages the moment as the recovery of an explicitly pagan historical episode, in a poem that is otherwise structured according to a uniquely Christian sacramental program, and it is Gower who constantly evokes the presence of Aristotle, "wys and expert in the sciences," and "that noble knyht" Alexander. Moreover Gower is careful to exclude from book 7 any post-Incarnation exempla (or even any mention of Christ's name, which makes it unique in the *Confessio*), turning the Christian Aristotelianism of his most immediate sources back toward its pagan origins.[52]

At one of the poem's most important moments of transition, then— when it sets aside both the amatory theme and the confessional structure of the first six books and openly embraces straightforward political instruction— the mechanism that enables that move is Amans's apparent curiosity about the pedagogical relationship between two virtuous pagans. This is in fact not a coincidence, though the narrative certainly tries to make it look like one, like an entirely contingent outgrowth of the lover's impulsive desires. Nor is it especially startling, however one might characterize the success of its integration—Gower is as much a medieval classicist as any of his contemporaries when it comes to valuing the "ethical expertise" of pagan folk and recognizing their role as an important source of theories of political organization and behavior perfectly consonant with his contemporary Christian world. But it is one more instance of the contemporary discourse of the virtuous pagan functioning as a formal resource for aesthetic organization. Like Chaucer, facing at the end of *Troilus and Criseyde* a difficult transition from courtly love to Christian prayer, Gower at this point in the *Confessio* finds in the convention of the righteous heathen the means to move from love to politics; if his gesture is less spectacular (and the results less compact) than Troilus's apotheosis, it is just as crucial to the structure of his poem, and just as dependent on the contemporary discourse of the virtuous pagan.

CONCLUSION

VIRTUOUS PAGANS AND VIRTUAL JEWS

Gower does tell one story in which the praiseworthy law of an unconverted pagan features prominently. This is the well-known tale of the Jew and the Pagan, which appears in book 7 in six manuscripts of Macaulay's second recension of the *Confessio amantis*. One summer, in the wilderness between Cairo and Babylon, two men fall in with one another by chance. As they travel and talk, one asks the other, "What man art thou, mi lieve brother? / Which is thi creance and thi feith?" (7.3220–21). As we have seen, such an inquiry into a stranger's "creance" puts us in virtuous pagan territory, and the traveler's answer confirms our expectation:

> "I am a paien," that other seith,
> "And be the lawe which I use
> I schal noght in mi feith refuse
> To loven alle men aliche,
> The povere both and ek the riche:
> Whan thei ben glade I schal be glad,
> And sori whan thei ben bestad;
> So schal I live in unite
> With every man in his degre.
> For riht as to myself I wolde,
> Riht so toward all othre I scholde
> Be gracious and debonaire."
>   (7.3222*–33*)

Like Mandeville's Brahmans and the *Polychronicon*'s Trajan, the Pagan follows a version of the Golden Rule.[1] His companion, however, obeys a different set of precepts; when asked about his beliefs, he replies:

> I am a Jew, and bi mi lawe
> I schal to noman be felawe

To kepe him trowthe in word ne dede,
Bot if he be withoute drede
A verrai Jew riht as am I;
For elles I mai trewely
Bereve him bothe lif and good.

(7.3239*–45*)

"Trewely" has a sly punning significance here, given the events that ensue.
The Jew, "which al untrowthe hadde" (7.3254*), asks to ride the Pagan's ass,
knowing that the Pagan's law enjoins exactly that kind of charity upon him;
when the Pagan agrees the Jew mounts up and rides off into the distance,
observing that "And in such wise as I thee tolde, / I thenke also for mi
partie / Upon the lawe of Juerie / To worche and do mi duete. / Thin asse
schal go forth with me / With al thi good, which I have sesed. . ."
(7.3290*–95*). Using another keyword of the virtuous pagan scene,
"rihtwisnesse," the dismayed Pagan prays to God—"O hihe sothfastnesse, /
That lovest alle rihtwisnesse, / Unto thi dom, lord, I appele" (7.3303*–04*)—
and at nightfall he is rewarded with the sight of the Jew "Al blodi ded
upon the gras, / Which strangled was of a leoun" (7.3322*–23*), the
unharmed ass wandering nearby. The sight prompts another brief prayer of
thanks.

Gower adapts this tale from the *Secretum secretorum* tradition, where it is
typically offered as a cautionary tale under the rubric of choosing an adviser
(e.g., in the Middle English translation entitled *The Governance of Lordschipes*
in MS Lambeth 501, dated soon after 1400[2]). Gower makes it an exemplum
of Pity, and in doing so alters the story in small but significant ways. In the
Secretum versions the Jew is not killed immediately but crippled when he is
thrown from the mule, and when the Pagan comes upon his injured antagonist he enacts the Good Samaritan paradigm, bearing him to his people,
where he later dies anyway. He thus demonstrates a thoroughgoing (if in the
circumstances understandably reluctant) devotion to his law—once burned,
he nevertheless shows mercy a second time. Moreover, the Pagan receives
material reward for his action: "And þe kyng of þat Citee whanne he herde
þe doynges of þat Enchanteour, he clepyd him afore him, and for his pity-
ous doynges, and for þe goodnesse of his lawe, he ordeyned him oon of his
Conseillers."[3] The story clearly cries out for an exegetical reading that,
in the secular context of the *Secretum*, it is necessarily denied. Gower, in
contrast, ends his narrative abruptly with the death of the Jew and turns to
the praise of pity, "thilke roote / Wherof the vertus springen alle" (7.3387*).
In the course of this passage he twice asserts that the story shows God's will
in action, defending and rewarding those who show pity: "And who this
tale redily / Remembre, as Aristotle it tolde, / He mai the will of god

beholde / Upon the point as it was ended, / Whereof that pite stod commended. . ." (7.3354*–58)—commended in this case by the king of the Heavenly City.

In thus modifying the tale for the *Confessio*, Gower actually makes it a somewhat more coherent exemplum; in the *Secretum*, it is not really a very good proof of the rule it purports to illustrate—i.e., "neuer haue trist yn man þat trowys noght þy lawe"—since the principle of discrimination embodied in that admonition actually governs the Jew in the tale, not the Pagan. By making the Pagan's law perfectly concordant with the finest version of Christian ethics in enjoining care and kindness for all peoples, Gower makes use of the triangulating ethnographic technique that François Hartog calls "the rule of the excluded middle," a strategy that permits a narrative trying to represent alterity to handle more than two terms (i.e, self and other) at once. The Pagan in his once-upon-a-time tale, set somewhere between the far away cities of Cairo and Babylon, is certainly an Other from the perspective of his Christian audience, but the Pagan's encounter with a figure who represents his own Other—the Jew, with his utterly contrary "lawe"—complicates any simple distinction between Pagan and Christian. In order to express adequately the alterity of the Jew, the Pagan is thus temporarily assimilated into the field of Christian ethics, and the middle term—a non-Christian but nevertheless morally admirable paganism—is essentially elided.[4]

There is of course another way to read this tale, not rhetorically or structurally but literally, and from this perspective its blunt message is "you can't trust Jews (but virtuous pagans are OK)." Such a reading is only encouraged by the use of a perfidious Jewish villain, a congenial stereotype for Christian writers throughout the Middle Ages; theoretically a person of any faith can fail to show pity, and book 7 offers up in quick succession the examples of Leontius, Siculus, Dionysius, Lichaon, and Spertachus, cruel tyrants all (*CA* 7.3267–3517). But in making an anti-ecumenical ruthlessness a supposed tenet of Jewish law and contrasting it with the laudable mercy of another non-Christian, this noxious little anecdote implies that the praise of pagan virtue is structurally dependent upon the denigration of Jewish vice.[5]

This conjunction is, unfortunately, the final piece in the structural analysis of the virtuous pagan scene in later Middle English treatments of the theme. It possesses an unpleasant but undeniable psychic logic, certainly; the anxiety generated by an affective investment in pagan righteousness, potentially threatening to the doctrinal integrity of Christianity, is relieved by its transformation into an anti-Judaic scorn that is then free to flow along the well-worn channels of medieval anti-Semitism. And the operation of this mechanism is visible in too many texts, and too many kinds of texts, to be a coincidental or even marginal phenomenon; it is clearly part and parcel of the period's interest in the figure of the righteous heathen.[6] Indeed, we have seen it already in the *Northern Homily Cycle*'s account of the Lost Tribes, where Alexander's prayer to move mountains is assigned one of the virtuous pagan scene's conventional morals: if a pagan can accomplish so much, what ought faithful Christians be able to do?

> Þat me ȝe seo · bi þis kyng
> Þat Wiþ good treuþe · gat his askyng
> And ȝit Was he heþene mon.
> And þerbi may We · seon vchon
> Þat cristene mon · ouhte muche more
> Gete · ȝif he in trouþe Wore
> Þen saraȝin · þat trouweþ nouht
> Þat Jhu Crist on Roode · haþ hym bouht.
>   (f.211r col.c)

Beyond this conventional Augustinian formula, though, is the same implicit structural point: God's grace shown to a pagan is intimately bound up with his continued punishment of the Jews, and Alexander's "treuþe," as I have argued, clearly derives in part from his disdain for the enclosed tribes; as he says to them, "Siþen ȝe han · oure god forsaken / And to maumetes · ȝe han [ȝ]ow taken / I · schal fonde · to sperre ȝow mare / Now ar euere · I · hennes fare." Moreover, again as we have already seen, the failure of this act of sublimation—now specifically legible as part of the virtuous pagan scene itself, and not just the more general border-policing endemic to Jewish–Christian relations in the Middle Ages—will ultimately have apocalyptic consequences.

> ȝit dwellen · mony Jewes þare
> And so schul þei · euer mare
> Til a ȝeyn · domes day.
> Þen schul þei alle · þeonnes stray
> In moni londes · schul þei Wende
> And Cristendom · schul þei schende.
> Ffor ouer al · þer þei go
> Cristene folk · schul þei slo.
> Þei schul leue · on Antecrist
> And Wene þat he beo · Ihu crist.
>   (f.211r col.c)

Similar sentiments and structures are visible in the version of the story contained in *Mandeville's Travels*. Sir John gives the most abbreviated account of the enclosure, basically omitting the reasons for it, but the most thorough (and tonally the most ambiguous)[7] description of the tribes' apocalyptic role, into which he enlists Jews living throughout the world:

> And yit natheles men seyn thei schulle gon out in the tyme of Antecrist, and that thei schulle maken gret slaughter of Cristene men. And therfore alle the Iewes that dwellen in alle londes lernen alleweys to speken Ebrew in hope that whan the other Iewes schulle gon out, that thei may vnderstonden hire speche and to leden hem into Cristendom for to destroye the Cristene peple. For the Iewes seyn that thei knowen wel be hire prophecyes that thei of Caspye schulle gon out and spreden thorghout alle the world, and that the Cristene men schulle ben vnder hire subieccoun als long as thei han ben in subieccoun of hem. (29.193)

The Hebrew language is thus no longer a scriptural language, the language of the Old Testament, nor even a cultural phenomenon, "the lettres that the Iewes vsen" (12.79); rather it is an instrument of international espionage, preserved not for the sake of ethnic identity but for anti-Christian antagonism. Of course, from the point of view of medieval anti-Semitism, such antagonism is largely what Jewish ethnic identity amounts to; in an earlier passage, one of the few in the *Travels* that considers the Jews collectively, they are accused, en masse, of seeking to "enpoysone alle Cristiantee" with an exotic venom.[8]

The incorporation of the Jews into the otherwise ecumenically generous world of *Mandeville's Travels* can thus only be imagined eschatologically, as several critics have noted,[9] and if as I have argued the virtuous pagan scene is what serves to structure the *Travels*, we should not be surprised—though we may still be dismayed—to find the Jews playing the same role as in the "Jew and the Pagan" story, writ apocalyptically large. "As unredeemable enemies, they secure the identity that always threatens to slip away from Mandeville's text," writes Greenblatt, but of course that identity is at issue in virtually every textual treatment of the righteous heathen, occasionally warranting the creation of a Jewish foil to enhance the brilliance of pagan virtue.[10] Such characters are of course "virtual" or "hermeneutical" Jews, like Lord Cobham's hypothetical heathen peoples; indeed, the repeated appearance of this phenomenon in various virtuous pagan scenarios helps further establish the ubiquity of the "hermeneutical Jew" in post-Expulsion England, that is, the conceptual creature who perfectly embodies Christian anxieties about the primacy of the Jewish covenant and the persistence of Jewish resistance to the revelation of the gospels, and whose various traits depend almost entirely on the rhetorical circumstances that call him into being.[11]

Virtuous pagan texts typically construct Jews who have a vexed relationship to religious law. Either Jewish law explicitly enjoins cruel and sinful behavior, as in the Jew and Pagan story, or the Jews themselves fail to live up to an otherwise praiseworthy law, as in the Lost Tribes tale. The *Northern Homily Cycle* specifies that the enclosed tribes were imprisoned "Ffor þei so ofte · god for sok / And to false maumetes · tok," and this reasoning is also advanced in another contemporary text that tells the tale, James Yonge's 1422 translation of the *Secretum, The Gouernance of Prynces*:

> Than [Alexander] enquerid wherfor thay were y-ladd out of har land, and he vnderstode by tham wych the verite knewen that for that thay weren into that traldome, that thay ne helde not the lawe of god of hevyn wyche thay had rescewid by Moyses, and wyrsepedyn fals goddis whych maket weryn by mannis handis; And therfore the prophetis of god prophiseden of hare thraldome, and Sayden that thay sholde not come agayn of that exil.[12]

Here and in the *Northern Homily Cycle* the Old Testament prophetic narrative, in which Israel is serially alienated and reconciled with God, is made into an incontrovertible law of history, and an implicit distinction is maintained between righteous Scriptural Hebrews—that is, Old Testament figures and patriarchs who observe the Mosaic law that medieval Christianity recognized as the proper Jewish faith—and historical "Jews," past, present, and future, who depart from that fixed and unchanging standard. Mandeville even enlists the Saracens in this conventional critique, early in the *Travels*: "Also the Sarazines sey that the Iewes ben cursed for thei han defouled the lawe that God sent hem be Moyses" (15.100).[13]

This concern with the law makes a certain sense as a strategy for addressing the supersessional anxieties that drove medieval thinking about Jewish–Christian relations; the need to preserve the idea of Mosaic law as a cherished heritage was necessarily accompanied by the need to demonstrate its inadequacy as a source of spiritual guidance in a post-Incarnation world—hence the distinction between Scriptural and historical Jews.[14] Virtuous pagan texts involving Jews complicate things by adding a third term to this neat binary. Adherence to pagan law, broadly defined, is represented as plainly virtuous through its contrast with Jewish behavior (either the depraved law of the Jew and Pagan story or the falling-off from Mosaic principles of the enclosed tribes); Gower's Pagan strives to "live in unite / With every man in his degre," while Alexander "Wiþ good treuþe · gat his askyng." At the same time, though, the moral adequacy of this pagan conduct implicitly challenges the Mosaic law that stands in the background of both of these stories, offering an alternative ethics that, while still inferior to the Christian scheme ("cristene mon · ouhte muche more / Gete. . ."), is clearly superior to the malice or backsliding of the Jews. In other words, there is more than one "middle" to be excluded here, and when virtuous pagans meet virtual Jews, the issue of supersession is transformed into a lineage-versus-patrimony conflict, in which Christian writers are drawn fictively to reject their Scriptural heritage in favor of a pagan ethics with an entirely different historical relation to their faith.

We have seen an analogous process at work in Gower's portrait of Alexander in the *Confessio amantis*, whom he represents as definitively rejecting the exotic, Egyptian heritage of his true father Nectanabus (by throwing him off a tower) and embracing instead the pragmatic and philosophical wisdom of a Greek father-figure, Aristotle—choosing his lineage rather than accepting a tainted patrimony. But the choice is even more fraught (and even more violently enacted) in the Middle English poem that most pointedly contrasts Jewish depravity and pagan, specifically Roman, virtue: *The Siege of Jerusalem.* The contrast is not, of course, a random one; the connection between the rise of the Empire and the spread of Christianity features largely in Christian historiography from Augustine onward, and the destruction of the Temple in Jerusalem in AD 70 is one of many historical episodes enlisted to support that interpretation of history. Versions of the story, complete with miraculous healings and conversions à la the tale of Constantine, circulated widely in encyclopedic texts like the *Golden Legend*, the *Speculum Historiale*, and the *Polychronicon* (and thus *Mandeville's Travels*[15]), as well as in shorter Latin and vernacular texts. In Middle English the most notorious version of the tale is *The Siege*, which survives in whole or part in nine manuscripts—a greater number than for any other alliterative poem but *Piers Plowman*.[16]

An account of the originary moment of what *Mandeville's Travels* calls the "subieccoun" of the Jews, the *Siege* begins when the Roman provincial governor Titus first hears the story of the gospel. When he curses the emperor whose appointment led to Pilate overseeing the Crucifixion, he is immediately cured of the disfiguring "canker" on his face; converted and baptized, he sets out to find a similarly miraculous remedy for his father Vespasian ("Waspasian" in Middle English), plagued since his youth by a "bikere of waspen bees" (34) in his nose. When his etymological malady is cured by means of an etymological relic—the vernicle of St. Veronica—the two noble Romans set out for Jerusalem with an armada to pass etymological judgment—"iewes"—upon the "Iewys" (270), who have not only refused to recognize Christ's divinity but have also refused to pay the required tribute to Rome. Vespasian, agent of a prophecy he has presumably not read (Matthew 24:2, "Amen, I say to you, there will not be left here a stone upon another stone that will not be thrown down"), vows outside the city that "Y to þe walles schal wende and walten alle ouere, / Schal no ston vpon ston stonde by Y passe" (355–56). The vow is ultimately fulfilled after a brutal siege, and the Jewish survivors are sold into slavery, thirty for a penny. Pilate commits suicide in prison.[17]

The Jews of the *Siege* are supposedly historical figures, the story of their defeat at the hands of recent Roman converts underwritten and authenticated by the power of the chronicle, but in fact their representation is just as fantastical, just as "hermeneutical," as Mandeville's apocalyptic Lost Tribes or the hypothetical Jew of the *Confessio amantis*. Discussing the poem's relation to the alliterative tradition, Ralph Hanna writes that the poet "describes the moment when the acts of the pagan worthies, classical and Jewish, were superseded and room left for that Christian heroic poetry he writes."[18] Indeed, the poem literally shows us the Jewish Worthies in the process of being replaced. When Caiaphas tries to inspire his forces with tales of their mighty ancestors—

Lered men of þe lawe	þat loude couþe synge
With sawters seten hym by	and þe psalmys told
Of douȝty Dauid þe kyng	and oþer dere storijs
Of [Iosue] þe noble Iewe,	and Iudas þe knyȝt.

(477–80)

—Vespasian responds by evoking the Passion, declaring that his army fights against "þis faiþles folke" (513) not for Nero's simple administrative quarrel but "Driȝten to serue" (521). But in this claim we can see that the parallel supersessions of classical and Jewish heroism are not identical processes— rather, Roman virtue is celebrated in the poem precisely because of its antagonism toward the Jews. Christine Chism observes that "The Jews are not the only threatening precursors in the poem; the poem pinions pagan Rome as well. It targets two pasts significant to Christendom as imperio-religious mythology, dispatching the rival religiosity of the Jews simultaneously with the crumbling imperial edifice of pagan Rome."[19] But pagan Rome is not so much dispatched as recuperated, in keeping with the motif that opposes heathen virtue to Jewish vice. Certainly not every Roman makes the grade; those who fail to move beyond an "unreconstructed paganism"—Nero, Galba, Otho Lucius, Vitellius, and Pilate—are assassinated or commit suicide.[20] But for them, at least, the potential exists to move from one category, "Roman and pagan," to another, "Roman and Christian"; such an opportunity is foreclosed to the Jews. Admittedly, the poem begins, after a brief account of the Passion, with the observation that God delays his vengeance on the Jews for forty years, ostensibly to see "ȝif þey torne wolde" (ll. 19–24), but this is a gesture in bad faith—of course they don't, and can't, if they are to remain the historical "Jews" of the medieval Christian imagination. Titus and Vespasian do, though, at the very first opportunity, a willingness that recalls the *facere* ethic and that right away marks the difference between the poem's totally unregenerate Jews and its potentially righteous Roman pagans, whose chivalric virtue needs only to be infused with the proper amount of crusading spirit.

The claim that the Middle English discourse of the righteous heathen was affiliated with the most venomous expressions of what Kathleen Biddick has called "the deeply tenacious melancholy of Christians for Jews"[21] makes a rather depressing conclusion to a study exploring the formal dimensions of a kind of ecumenical thinking not typically associated with the writing of the Middle Ages. It will not do to see the former as completely absorbed by the latter, to be sure; righteous heathen stories of the Trajan/Erkenwald type, specifically focused on salvation, tend not to include any explicitly anti-Jewish structural elements. But there can be little question that *The Siege of Jerusalem* darkly mirrors the Gregory/Trajan legend, which is also at its root an attempt to reclaim a classical lineage from beyond Christianity's doctrinal border; the *Siege*, in fact, miraculously rescues not one but two Roman emperors, though in thus recuperating one of Christianity's losses it violently disavows another, equally potent one.[22]

One thing that this aspect of the topic does reveal, however, is the insistent reach of the analogy with which I began this book, between medieval texts about virtuous pagans and modern attitudes toward our medieval inheritance. Acknowledging the connection between representations of pagan virtue and concomitant condemnations of hypothetical Jewish vice reveals a reflex of the medieval Christianist master narrative that we have presumably seen through nowadays, whether or not we are part of Jill Mann's cadre of atheist critics, and one that we have chosen to reject as either an ethical or a critical practice. And what remains after our own disavowals, inevitably, is form. And thus I will end not with the complicity of righteous heathen discourse in medieval anti-Semitism, but with another modern instance, one that, paradoxically, testifies to this persistence of form precisely in the way that it registers the difference between virtuous pagan stories then and now.

Time Magazine's cover story for the October 25, 1999 issue was entitled "A Week in the Life of a High School," and it chronicled, in a generous fifty pages, a week's events at the public high school in Webster Groves, Missouri, a tree-lined, "bellwether" suburb of St. Louis with a largely Catholic population of about 23,000 (St. Louis, a city named for a pious crusader-king, is where I write these final pages). Among the vignettes describing sports and pep rallies, parties and property tax votes, and post-Columbine school security was the story of Ben Averbuch, a fifty-eight-year-old school counselor who was also the girls' softball coach, and who died unexpectedly of a stroke in the middle of the season.

> Jewish custom does not include flowers at a burial. But the rabbi at Ben Averbuch's made an exception, and one by one, the softball team dropped yellow roses into his grave. "I think his family was amazed at how many people were there," says Meg [Kassabaum, a reserve third baseman].

Immediately after the death, people asked Bob Berndt [another coach] if the Stateswomen were canceling the rest of their schedule. "Coach wouldn't want us to," he replied. But some things did change. After warming up and just before breaking out of their huddle, the girls now recite a Hail Mary. Then they yell, "Averbuch, pray for us!"

Becca [Dunn, the centerfielder] has considered the appropriateness of this. "I wanted him to be in heaven so much," she says. "But I didn't know for sure. But I was reading the Bible—*Romans 10*. And it explained that God asks the Jews to come live with him based on faith. He chooses who he wants to be in heaven. So you just think Mr. Averbuch is in heaven." And the girls play with him in mind.

"He was definitely that kind of guy," says Meg. "You know we're winning it for him."[23]

The conventional connection between Jewish conversion or salvation and the Virgin Mary, here somewhat attenuated, is only one of the medieval aspects of this story, and if it offers a rather liberal reading of Romans 10, we can see in the pronoun slippage that elides what God does ("He chooses who he wants to be in heaven") and what Becca Dunn does ("So you just think Mr. Averbuch is in heaven") a rhetorical gesture typical of the virtuous pagan scene. And this is certainly another exceptional salvation story, in which the girls are moved to override the doctrinal differences that divided the largely Christian team from its Jewish coach because of their affection and esteem for a man whose secular pastoral role involved, among other things, college placement and finding people scholarships—helping the "lewed" become "letrede." "Wheiþer he be saaf or noȝt saaf, þe soþe woot no clergie," but for those who believe in the possibility of such things as exceptional salvations, open to all those who "live in unite / With every man in his degre," he was, it seems, definitely that kind of guy.

NOTES

Introduction   The Rule of Exceptional Salvations

1. Jacobus de Voragine, *The Golden Legend*, trans. William Granger Ryan, 2 vols. (Princeton: Princeton University Press, 1995), I.178–79. For the Latin text, see the edition of Theodor Graesse, *Legenda Aurea Vulgo Historia Lombardica Dicta*, 2nd edn. (Leipzig: Librariae Arnoldianae, 1850), pp. 196–97.

2. At B.11.160–61: "This matere is merk for many, ac men of holy chirche, / The *legenda sanctorum* yow lereþ more largere þan I yow telle." All quotations from *Piers Plowman* B derive from the Athlone edition, ed. George Kane and E. Talbot Donaldson (London: Athlone Press, 1975). On the popularity of the *Legenda*, see the introduction to Ryan's edition, pp. xiii-xiv.

3. Gordon Whatley, "The Uses of Hagiography: The Legend of Pope Gregory and the Emperor Trajan in the Middle Ages," *Viator* 15 (1984): 25.

4. As in the case of Gregory himself, for example: "The name Gregory (Gregorius) is formed from *grex*, flock, and *gore*, which means to preach or to say, and Saint Gregory was preacher to his flock. Or the name resembles *egregarius*, from *egregius*, outstanding, and *gore*; and Gregory was an outstanding preacher and doctor. Or Gregorius, in our language, suggests vigilance, watchfulness; and the saint watched over himself, over God, and over his flock. . ." de Voragine, *Golden Legend*, p. 171.

5. de Voragine, *Golden Legend*, p. 179.

6. This conclusion to the story is ruled as well by a logic of substitution: not only is Gregory's original emotional anguish ("bitter tears") replaced by physical pain, but Trajan's now-remediated suffering (however imagined) is also transferred to Gregory, albeit in earthly form. Of course, Jacobus's attempt to domesticate the Trajan story is at the same time embarrassed by this substitution; Gregory's suffering leads us away from Trajan's on the level of narrative, but it also points back to it, and to the question of justice *sub conditione*. Gregory's own salvation is of course implied by the mention of time spent in purgatory, but how are we to measure that against time spent in hell?

    Quoting from Gregory's letters may have been Jacobus's own idea; while the link between the intervention and Gregory's well-known infirmities is established very early in the tradition, none of the earlier accounts that I have seen use the letters to reinforce the connection. See Whatley, "Uses," p.36n.39. The letters survive in almost a hundred manuscripts; they have recently been newly translated by John R.C. Martyn (*The Letters of Gregory the Great*, 3 vols., Mediaeval Sources in Translation 40 [Toronto: Pontifical Institute of Mediaeval Studies, 2004]). The letters quoted in the *Golden Legend* are 9.232 and 10.14 in Martyn's edition.

7. Quoted in Theresa Coletti, *Naming the Rose: Eco, Medieval Signs, and Modern Theory* (Ithaca, NY: Cornell University Press, 1988), p. 8, from Walter E. Stephens, "Ec[h]o in Fabula," *Diacritics* 13 (1983): 51–64.

8. For the earliest version of the story, see Whatley, "Uses," pp. 27–29.

9. On Uhtred de Boldon, see Mildred Elizabeth Marcett, *Uhtred de Boldon, Friar William Jordan, and Piers Plowman* (New York: privately printed, 1938); M.D. Knowles, "The Censured Opinions of Uthred of Boldon," *PBA* 37 (1951): 305–42; and G.H. Russell, "The Salvation of the Heathen: The Exploration of a Theme in *Piers Plowman*," *JWCI* 29 (1966): 101–116. Nicholas Watson discusses the idea of universal salvation in Middle English texts in "Visions of Inclusion: Universal Salvation and Vernacular Theology in Pre-Reformation England," *JMEMS* 27 (1997): 145–87.

10. Thomas Aquinas, *Summa Theologica* III, suppl., q. 71 a. 5, obj. 5 and resp. ob. 5, in *Opera Omnia*, 25 vols. (Parma: Petri Fiaccadori, 1852–73; repr. NY: Musurgia Publishers, 1948), 4: 588–89. I present here the Dominican translation: *Summa Theologica*, trans. Fathers of the English Dominican Province, 22 vols. (London: Burns, Oates and Washbourne, Ltd., 1921–24), 20: 53–4.

    For a survey of scholastic theologians' treatment of Trajan and stories of his salvation, including Bonaventure, William of Auxerre, and Alexander of Hales, see Whatley's "Uses," especially pp. 35–39; Louis Capéran, *Le problème du salut des infidèles: essai historique*, 2nd edn. (Toulouse: Grand Seminaire, 1934), pp. 186–218; Marcia L. Colish, "The Virtuous Pagan: Dante and the Christian Tradition," in *The Unbounded Community: Papers in Christian Ecumenism in Honor of Jaroslav Pelikan*, ed. William Caferro and Duncan G. Fisher (New York: Garland, 1996), pp. 43–92; and the account in Cindy L. Vitto, "The Virtuous Pagan in Middle English Literature," *Transactions of the American Philosophical Society* 79 (1989), chap. 3.

11. Both questions are part of Q.71, which is concerned with "suffrages for the dead." This caution against turning Trajan's case into a general rule essentially repeats Aquinas's remarks in his *Commmentum in quattuor libros sententiarum* I, dist. 43, q. 2 a. 2, obj. 5 and ad. 5; see *Opera Omnia* 6.351–52. Whatley ("Uses," pp. 39–40) argues that Aquinas's "final word" on the subject represented a conservative retreat "from the resuscitation argument that he and other schoolmen consistently propounded elsewhere."

12. As Marcia Colish observes, for Aquinas "Trajan is not a unique case but a member of an entire cohort of souls who are 'not finally damned' (*non finaliter damnati*)," but "[h]ow and why Trajan and the other members of this cohort get classified incorrectly in the first place and therefore endure centuries of undeserved punishment in hell is a question that Aquinas neither raises nor answers." Colish, "The Virtuous Pagan," p. 68.

13. Aquinas, *Summa Theologica* 2a2ae q. 2 a. 7, obj. 3 and resp. ob. 3: "Whether explicit faith in the Incarnation is necessary for salvation?"

14. Useful surveys of this topic include Capéran's magisterial *Le problème du salut des infidèles*; Colish, "The Virtuous Pagan"; Vitto, "The Virtuous Pagan in Middle English Literature"; Gordon Leff, *Bradwardine and the Pelagians* (Cambridge: Cambridge University Press, 1957); Heiko Obermann, "*Facientibus quod in se est Deus non denegat gratiam*: Robert Holcot, O.P., and the Beginning of Luther's Theology," *Harvard Theological Review* 55 (1962): 317–42; and Alastair Minnis, *Chaucer and Pagan Antiquity* (Cambridge: D.S. Brewer, 1982).

15. The Gregory–Trajan legend may have originated in England; the story first appears in the earliest life of Gregory, written by an anonymous monk of Whitby sometime in the eighth century. See *The Earliest Life of Gregory the Great*, ed. and trans. Bertram Colgrave (Lawrence, KS: University of Kansas Press, 1968).

    For the coinage "vernacular theology," see Nicholas Watson, "Censorship and Cultural Change in Late-Medieval England: Vernacular Theology, the Oxford Translation Debate, and Arundel's Constitutions of 1409," *Speculum* 70 (1995): 822–64, esp. pp. 823–24 and 823n.4.

16. For a survey of the criticism of Langland's use of the righteous heathen, see chapter one. On the habit of imitation, Watson observes that "Historical scholarship—the kind that seeks to build stories out of textual and material remains and even the kind that critiques this storytelling—tends to find itself paraphrasing or repeating the past, as its language and assumptions are pulled magnetically toward those of the subject under discussion." See his "Desire for the Past," *SAC* 21 (1999): 59–97, here p. 91.

17. For the provocative and parallel claim that "writing in the vernacular *exerts pressure* toward a universalist understanding of the meaning of the Incarnation as an expression of the illimitability of divine love," see Watson, "Visions of Inclusion: Universal Salvation and Vernacular Theology in Pre-Reformation England," *JMEMS* 27 (1997): p. 171. Emphasis in original.

18. For a vigorous and polemical defense of "reading for form," see Ellen Rooney, "Form and Contentment," *Modern Language Quarterly* 61 (2000): 17–40; she writes, "When the text-to-be-read (whatever its genre) is engaged only to confirm the prior insights of a theoretical problematic, reading is reduced to reiteration and becomes quite literally beside the point. One might say that we overlook most of the work of any text if the only formal feature we can discern in it is a reflected theme, the mirror image of a theory that is, by comparison to the belated and all-too-predictable text, seen as all-knowing and, just as important, as complete" (29–30).

    In exploring the change wrought on theological topics when they are absorbed into specifically literary vernacular discourse, my project is analogous to the recent work of Jim Rhodes, *Poetry Does Theology: Chaucer, Grosseteste, and the Pearl-Poet* (Notre Dame, IN: Notre Dame University Press, 2001). Rhodes's study is broader in scope than mine, but we are both interested in "What happens. . .when poetry deals explicitly with a serious theological issue" (2) and in "reading the poems as fictions and not as an armature for theology" (10); he approvingly quotes John Burrow on *Piers Plowman's* Christ, who "becomes, for the purposes of the poem, subject to the laws of fiction and the exigencies of art" (5), an admirably succinct expression of the concept.

19. Hayden White, "The Value of Narrativity in the Representation of Reality," in *The Content of the Form: Narrative Discourse and Historical Representation* (Baltimore: Johns Hopkins University Press, 1987), p. 2.

20. Capéran, *Le problème*, pp. 9–70, discusses treatments of the issue from the gospels through the early church fathers. As a solution, of course, the salvation of the heathen creates its own set of problems, not to mention two millennia of speculation about how to solve them, but it is worth noting that issues of method in particular—how exactly might pagans be saved?—are secondary problems, derivative of the desire to save them in the first place.

21. "Televangelist Larry Lea's Followers Speak," *San Francisco Chronicle/Examiner*, Sunday, November 18, 1990. Lea was trained at Oral Roberts University and was the founder of International Prayer Ministries.

22. "Televangelist Larry Lea's Followers Speak": "As we entered the Civic Center the protesters were kicking and spitting and I thought, 'I knew it! We're going to become martyrs!' Someone screamed that I was a Christian bigot. That's a lie! I'm so angry, the way they were judging me. They have no concept of why I serve God, why I came here." The evocation of an apostolic martyrdom in the midst of a "crusade" is of course one more thing that associates this episode with conventional medieval tales of the conversion of the heathen.

23. All quotations from *Mandeville's Travels* are taken from M.C. Seymour's edition of the Cotton text (Oxford, 1967), and will be cited parenthetically by chapter and page number, here 32.215.

24. That indebtedness is captured in almost epigrammatical form by Langland's Imaginatif, speaking of Aristotle et al.: "For letrede men were lewede yet, ne were loore of hir bokes" (B.12.276).

25. Jill Mann, "Chaucer and Atheism," *SAC* 17 (1995): 5–19, here p. 5. Mann goes on to argue that "dialogue" is the key to addressing the question of difference or rupture, and that it offers a potential escape from a "falsifying historicism" on the one hand and a contemporary solipsistic subjectivism on the other, "because it acknowledges that the participants in the dialogue are from the outset in a situation with respect to each other and that this situation will condition both the form that utterance takes and the way it is understood" (8). She concludes, then, with what we might call the weak or pragmatic version of antifoundationalism associated with Rorty and Fish. Dialogue, of course, is exactly what Middle English writers struck on too; rather than simply relating recuperative narratives of virtuous pagans, they produced narrativized dialogues in which pagan folk apparently spoke for themselves, in what medieval readers were encouraged to take as the semblance of their "real" voices. What modern criticism does with medieval texts and textuality is clearly analogous, and while there are limits to the

insights that thinking by analogy can produce, there are certain things that it can reveal quite sharply.

26. The phrase is Foucault's, from the 1969 essay "What is an author?"; I quote from the translation by Josué V. Harart, reprinted in *Contemporary Literary Criticism: Literary and Cultural Studies*, ed. Robert con Davis and Ronald Schleifer, 4th edn. (New York: Longman, 1998), p. 372.

27. For provocative remarks about the role of desire in modern medieval studies see L.O. Aranye Fradenburg, *Sacrifice Your Love: Psychoanalysis, Historicism, Chaucer* (Minneapolis: University of Minnesota Press, 2002), pp. 239–52, and the earlier version of this chapter in " 'So that we may speak of them': Enjoying the Middle Ages," *New Literary History* 28 (1997): 205–30.

28. On medieval studies as a clerisy, see Lee Patterson, "On the Margin: Postmodernism, Ironic History, and Medieval Studies," *Speculum* 65 (1990): 87–108, and "Introduction," in *Literary Practice and Social Change in Britain, 1380–1530*, ed. Patterson, (Berkeley and Los Angeles: University of California Press, 1990), p. 3.

29. Stephen Greenblatt, *Shakespearean Negotiations: The Circulation of Social Energy in Renaissance England* (Berkeley: University of California Press, 1988), p. 1.

30. Certainly the Middle Ages was less embarrassed and troubled by this problem than modern criticism; thus one could legitimately compare medieval treatments of the virtuous heathen to the more totalizing exegetical historicism conventionally associated with D.W. Robertson, insofar as saved pagans were always recruited into the Christian view of the cosmos rather than allowed to go their own way (since there was really nowhere else to go). Such an approach doesn't do away with the formal paradoxes at the heart of the matter, though; it simply declares them to be irrelevant in the end.

31. For a relatively early example of such a critique see Joel Fineman, "The History of the Anecdote: Fiction and Fiction," in *The New Historicism*, ed. H. Aram Veeser (London: Routledge, 1989), pp. 49–76.

32. Watson, "Desire for the Past," p. 87. The unpredictability of such empathy, and the emotional power of its claims, are once again nicely captured in the *Golden Legend*, where "one day" (L: *quadam vice*) Gregory is reminded of Trajan's story and it moves him to tears. Watson writes that "the past will always push through in this emotion-laden way in our accounts of it—isn't this in fact the very point made over and over again in Toni Morrison's *Beloved?*—a story that passes itself on into the future whether or not it is the story we want to be telling" ("Desire," p. 72). The distance from the *Golden Legend* to Morrison's *Beloved* is not so great as it seems; Watson uses the example in the course of rebutting Kathleen Biddick's critique (in *The Shock of Medievalism*) of Caroline Walker Bynum's "empathetic approach to the past" (p. 60) in *Holy Feast and Holy Fast*. His essay stresses "the need to think clearly about the way all such study has emotional designs on its object" (p. 61), and he acknowledges Fradenburg's psychoanalytic work on this topic (see n.27 above) while describing his own efforts as "mystical" (that is, concerned with affect and empathy as experienced and described by medieval

mystics). For a more broadly historicist approach to the affective dimension of medieval studies we can turn to Stephanie Trigg's *Congenial Souls: Reading Chaucer from Medieval to Postmodern* (Minneapolis: University of Minnesota Press, 2002), which takes up "the discourses of affinity. . .by which Chaucer's readers write themselves into relationships of intimacy with the poet and his various reading communities" (pp. 4–5). See especially her remarks on the role of "voice," which implies "the possibility of hearing Chaucer speak across the centuries and across different cultures" and "becomes a inclusive and enabling trope that allows us to contain and restrict difference. . ." (p. 22). Chaucer has been singled out since the sixteenth century by successive generations of readers and critics as a voice that must be recovered despite difference, distance, and changes in "forme of speche"; he is modern literary criticism's virtuous pagan, a Trajan figure whose commemoration continually requires us (that is, all us medievalists) to adopt Gregory's role.

33. On the difference between "reading for form" and an aestheticizing formalism, see Susan J. Wolfson, "Reading for Form," *Modern Language Quarterly* 61 (2000): 1–16.

34. Lee Patterson, *Negotiating the Past: The Historical Understanding of Medieval Literature* (Madison, WI: University of Wisconsin Press, 1987), pp. 210–22.

35. On the sometimes constrained nature of these interfaith dialogues—specifically fictionalized Jewish/Christian exchanges—see Steven F. Kruger, "The Spectral Jew," *NML* 2 (1998): 20–28. He writes that "Presenting Jewish voices and figures in such works involves a certain embodying of the Jew: Jewish presences are conjured up 'to speak for themselves.' But of course, they speak as their Christian authors determine, and their embodiment usually remains sketchy and their voices weak" (p. 21). The voices of virtuous pagans are of course equally the invention of Christian authors; my argument throughout the book, though, is that they are anything but weak.

Chapter 1   The Trouble with Trajan

1. See Introduction, n.15 above, and Gordon Whatley, "The Uses of Hagiography: The Legend of Pope Gregory and the Emperor Trajan in the Middle Ages," *Viator* 15 (1984): 27–29.

2. Nancy Vickers, "Seeing Is Believing: Gregory, Trajan, and Dante's Art," *Dante Studies* 101 (1983): 74.

3. Frank Grady, "*Piers Plowman*, *St. Erkenwald*, and the Rule of Exceptional Salvations," *YLS* 6 (1992): 61–86.

4. There is, possibly, a loose allegorical significance to the widow story, if we read the innocent, dead son as referring to Christ and the widow as a figure for the Church; in this case Trajan's action is a facsimile of a conversion, or at least a service rendered to the Church. But I know of no such medieval glosses.

5. Ranulph Higden, *Polychronicon Ranulphi Higden monachi Cestrensis*, ed. J.R. Lumby, 9 vols., Rolls Series 41 (London: Longman and Co., 1865–86), 5:5. Higden cites "Helymandus" (i.e., Helinand of Froidmont, one of Vincent of Beauvais's sources) for his Trajan material; according to Bunt (1986,

139n10), he may have taken it directly from Vincent, who also cites Helinand at X.46. Helinand's Trajan material is now lost. Vincent does not mention Trajan's salvation there but in X.48, where he also quotes Eutropius on the "golden rule" stanza. Another Middle English text that cites Helinandus also places the widow story and the miraculous salvation in different chapters; this is the fifteenth-century *Alphabet of Tales*, a translation of the Dominican *Alphabetum narrationum* (ca. 1300?), which tells two versions of the widow story (CCCCX, "Iudex bonus. . ."; CCCCXIX, "Iustitia. . .") and later provides a condensed account of Gregory's intervention (DXCII, "Oracione reuocatur ab inferis dampnatus"). See Mary Macleod Banks, ed., *An Alphabet of Tales*, EETS o.s. 126–27 (London: Kegan Paul, Trench, Trübner and Co., 1904–5), 2.281–82, 287, 393.

6. Higden, *Polychronicon*, 5:7.

7. Higden's observation is unintentionally ironic: the legend suggests that Gregory was inspired by one of the reliefs in the Forum of Trajan, and Gaston Paris, Vickers, and others have speculated that the source of the story of the widow was probably just such a sculpture, depicting Trajan on horseback receiving obeisance from a kneeling female figure, allegorically representative of a conquered province. Such a figure is visible today on Trajan's column in Rome (Vickers, p. 76). Misinterpreted, this piece would become a depiction of the poor widow grieving for her murdered son. The story of the widow apparently stems from an incident involving Hadrian, only later conflated with the Forum sculpture; thus Higden reverses the likely order by making the episode of the widow a source of the monument. See Gaston Paris, "La Légende de Trajan," *Mélanges de l'école des hautes études*, Fasc. 35e (Paris: École pratique des hautes études, 1878), pp. 289–95; Whatley, "Uses," p. 25n.2; and *St. Erkenwald*, ed. Sir Israel Gollancz (Oxford: Oxford University Press, 1922), p. xl.

8. Higden, *Polychronicon*, 5:7. Trevisa refrains from editorializing when the topic comes up again in book 5: "mynde of þat myracle of þe deliveraunce of Traianus at þe sepulcre of þe apostles in þe citee of Rome, by þe grete Gregorie pope, is i-holde, as everich ȝere holy chirche makeþ mynde" (Higden, *Polychronicon*, 6:197). Higden does not supply a date for this annual observance, whose source may be a mistranslated or misremembered passage in Gerald of Wales's *Liber de principis instructione*, dist. I c.xviii (which Higden cites as c. xvii), describing an instance of secular commemoration; see *Giraldi Cambrensis Opera*, ed. J.S. Brewer, J.F. Dimoch, and G.F. Warner, 8 vols., Rolls Series 21 (London: Longman, Green, Longman and Roberts, et al., 1861–91), 8:16–17, 82–84.

9. On the mid-century "Pelagian controversy," see Gordon Leff, *Bradwardine and the Pelagians* (Cambridge: Cambridge University Press, 1957). On Uhtred see Marcett, *Uhtred de Boldon, Friar William Jordan, and Piers Plowman* (New York: privately printed, 1938); M.D. Knowles, "The Censured Opinions of Uhtred of Boldon," *PBA* 37 (1951): 305–42; and G.H. Russell, "The Salvation of the Heathen: The Exploration of a Theme in *Piers Plowman*," *JWCI* 29 (1966): 101–16. For Hilton, see Watson, "Visions of Inclusion: Universal Salvation and Vernacular Theology in Pre-Reformation England," *JMEMS* 27

(1997): 146–48 (my source for the phrase "deviant views of redemption"), and A.J. Minnis, "Looking for a Sign: The Quest for Nominalism in Chaucer and Langland," in *Essays on Ricardian Literature in Honour of J.A. Burrow*, ed. Minnis, Charlotte C. Morse, and Thorlac Turville-Petre (Oxford: Clarendon Press, 1997), pp. 142–44.

10. Whatley, "Uses," pp. 26–27.

11. R.W. Chambers, "Long Will, Dante, and the Righteous Heathen," *Essays and Studies* IX (1923): 76.

12. T.P. Dunning, "Langland and the Salvation of the Heathen," *MAE* (1943): 46.

13. Critics asserting Langland's general Augustinianism include Elizabeth Doxsee, " 'Trew Treuth' and Canon Law: The Orthodoxy of Trajan's Salvation in *Piers Plowman* C-Text," *Neuphilologische mitteilungen* 89 (1988): 295–311; Denise Baker, "From Plowing to Penitence: *Piers Plowman* and Fourteenth-Century Theology," *Speculum* 55 (1980): 715–25; and Rosemary Woolf, "The Tearing of the Pardon," in *Piers Plowman: Critical Approaches*, ed. S.S. Hussey (London: Methuen, 1969): 50–75. The liberal side includes Marcett, *Uhtred de Boldon*; Russell, "Salvation of the Heathen"; Watson, "Visions of Inclusion"; Coleman, *Piers Plowman and the Moderni* (Rome: Edizioni di storia e letteratura, 1981); Robert Adams, "Piers's Pardon and Langland's Semi-Pelagianism," *Traditio* 39 (1983): 367–418; Pamela Gradon, "*Trajanus Redivivus*: Another Look at Trajan in *Piers Plowman*," in *Middle English Studies Presented to Norman Davis in Honour of His Seventieth Birthday*, ed. Douglas Gray and E.G. Stanley (Oxford: Clarendon Press, 1983): 93–114; Gordon Whatley, "*Piers Plowman* B 12.277–94: Notes on Language, Text, Theology," *Modern Philology* 82 (1984): 1–12; and R.F. Green, *A Crisis of Truth: Literature and Law in Ricardian England* (Philadelphia: University of Pennsylvania Press, 1999), pp. 368–76. Cindy L. Vitto, in "The Virtuous Pagan in Middle English Literature," *Transactions of the American Philosophical Society* 79 (1989), stakes out an uncontroversial middle ground, concluding that "[*Piers* and *St. Erkenwald*] nicely balance each other on the issue of grace (or baptism) vs. works: neither alone will suffice for salvation" (p. 89).

14. Robert Adams, "Langland's Theology," in *A Companion to Piers Plowman*, ed. John A. Alford (Berkeley: University of California Press, 1988), esp. pp. 95–98.

15. Minnis, "Looking for a Sign," p. 177 n.83.

16. Adams, "Langland's Theology," p. 96.

17. R.W. Chambers, *Man's Unconquerable Mind* (London: J. Cape, 1939), p. 148. Cf. also his remark in "Long Will" (53) on the pardon scene: "When the priest, representing current ideas, refuses to accept it, the poet is brought up against the contrast which he feels so bitterly, between his own sense of justice, and that which seems to him to prevail in the current practice of the church."

18. Chambers, "Long Will, 64. On Langland and the *moderni*, see Coleman, *Piers Plowman and the Moderni*, pp. 17–35, and Adams, "Langland's Theology," pp. 107–09.

19. Minnis, "Looking for a Sign," p. 145.

20. Minnis, "Looking for a Sign," p. 156. Compare Pamela Gradon, "*Trajanus*," pp. 103ff: she also sees no evidence of explicit nominalism, but nevertheless adheres to the semi-Pelagian line.

21. Lee Patterson, "Historical Criticism and the Development of Chaucer Studies," in his *Negotiating the Past: The Historical Understanding of Medieval Literature* (Madison, WI: University of Wisconsin Press, 1987), pp. 3–39. It should be noted that critics writing about Trajan have tended to cite more widely from scholastic theology than from the patristic sources preferred by Robertson and Huppé, a habit they share with later practitioners of exegetical criticism; see Anne Middleton, "Introduction: The Critical Heritage," in *A Companion to Piers Plowman*, pp. 17–18.

22. Middleton, "Critical Heritage," pp. 15–16.

23. Chambers, *Man's Unconquerable Mind*, p. 147. Trigg's "discourses of affinity" are apparently operative in the reading of Langland as well as in the reading of Chaucer; see Introduction, n.32.

24. For example, the obvious overstatement that "Langland recombines the two parts of the story in a way that effectively robs Gregory's act of any efficacy whatever"—which appears in my "Rule of Exceptional Salvations," p. 70. Mistakes were made.

25. Whatley, "*Piers Plowman*," pp. 10–11; Minnis, "Looking for a Sign," pp. 165–66.

26. E.g. in Whatley, "*Piers Plowman*," an essay entirely devoted to the difficulties of this passage.

27. Watson, "Visions," pp. 157–60; Minnis, "Looking for a Sign," 159–62; Russell, "Salvation of the Heathen."

28. ". . .one could point to the sudden appearance of St. Paul in Chaucer's 'Second Nun's Tale,' momentarily *present* (literally, present*ed*) to deliver a text from Ephesians, and to similar moments in *Piers Plowman*, in which an authority figure, like Piers himself or Trajan, abruptly materializes to deliver a brief text. In all instances the text could simply have been quoted or cited, but for the medieval reader or visualizer, there was an obvious (and perhaps ancient) satisfaction and mnemonic clarity in having it literally 'represented' and orally delivered." Kathryn Kerby-Fulton and Denise L. Despres, *Iconography and the Professional Reader: The Politics of Book Production in the Douce Piers Plowman* (Minneapolis: University of Minnesota Press, 1999), p. 104.

29. Joseph Wittig surveys the various opinions in his " 'Piers Plowman' B, Passus IX–XII: Elements in the Design of the Inward Journey," *Traditio* 28 (1972): 255 n. 143. James Simpson (*Piers Plowman: An Introduction to the B-Text* [London: Longmans, 1990], p. 127) argues that Trajan's speech extends to line 317, giving him over 150 lines; Kane and Donaldson's edition attribute a mere 15 lines to Trajan.

30. It also puts in context some of the statements made about Trajan, for example Chambers's well-known asseveration that Langland "contradicts all known authorities in making Trajan's salvation depend solely upon his own virtues," ("Long Will," p. 66)—a statement that we might see as being made under the influence of Trajan's own vigorous personality. But cf. Simpson (126–28), on why Trajan's speech should not be seen as "presumptuous."

31. Wittig, "Elements," p. 280. On the cognitive structure of the dream, see also Simpson, *Piers Plowman*, pp. 91–107. On the signature passages in the poem and in this vision, see Anne Middleton, "William Langland's 'Kynde Name': Authorial Signature and Social Identity in Late Fourteenth-Century England," in *Literary Practice and Social Change in Britain, 1380–1530*, ed. Lee Patterson (Berkeley: University of California Press, 1990), pp. 15–82.

32. This observation presumes the canonical but by no means universal understanding of the poem's composition history, A to B to C; see Kathryn Kerby-Fulton, "Piers Plowman," in *The Cambridge History of Medieval English Literature*, ed. David Wallace (Cambridge: Cambridge University Press, 1999), p. 520; and Grady, "Chaucer Reading Langland: *The House of Fame*," *SAC* 18 (1996), pp. 13–14. The importance of the virtuous pagan in linking one phase of the poem to the next offers an interesting analogy to *Mandeville's Travels*, where Sir John's conversation with the Sultan marks the point of connection between the texts' two main sources, Boldensele's *Itinerarius* and Odoric's *Relatio*. See chapter 2.

33. Simpson, *Piers Plowman*, p. 124.

34. Anne Middleton, "Narration and the Invention of Experience: Episodic Form in *Piers Plowman*," in *The Wisdom of Poetry: Essays in Early English Literature in Honor of Morton W. Bloomfield*, ed. Larry D. Benson and Siegfried Wenzel (Kalamazoo, MI: Medieval Institute Publications, 1982), pp. 91–122, 280–83.

35. Wittig, "Elements," p. 258.

36. The consideration at the end of passus 10 of hard cases like Mary Magdalene and the good thief does not specifically introduce the issue on the salvation of non-Christians; they are instances of counterintuitive salvations, not exceptional ones.

   On Langland and the Jews, and specifically his evolving account of the relationship of Judaism to Christianity, see Elisa Narin van Court, "The Hermeneutics of Supersession: The Revisions of the Jews from the B to the C text of *Piers Plowman*," *YLS* 10 (1996): 43–88.

37. Compare Mandeville's brief lament (XV.103) about the Christians under Saracen rule who have converted to Islam, an act that he finds reprehensible but not theoretically impossible.

38. Simpson, *Piers Plowman*, pp. 128–36.

39. Trajan's vexed relationship to the faculty model can be seen in the pattern of citations in Wittig's seminal essay, "Elements of Design"; this richly annotated piece quotes impressively and to great effect from the tradition of monastic "moral" psychology while examining the portions of the third vision that precede and follow Trajan's appearance, but while discussing Trajan himself can only quote from the various sources and commentaries on the Gregory/Trajan legend, since Trajan never figures explicitly (or implicitly) in the monastic literature.

40. The situation is different, of course, in texts operating under different generic rules, like travel literature (e.g., *Mandeville's Travels*) or romance history (e.g., the Alexander–Dindimus exchange).

41. On Dante's Cato, and the salvation of the heathen in *The Divine Comedy* generally, see Kenelm Foster, *The Two Dantes and other studies* (London: Darton, Longman and Todd, 1977); David Foster, "Dante's Virtuous Pagans," *Dante Studies* 96 (1973): 145–62; and Gino Rizzo, "Dante and the Virtuous Pagans," in *A Dante Symposium, in Commemoration of the 700th Anniversary of the Poet's Birth (1265–1965)*, ed. William De Sua and Gino Rizzo, University of North Carolina Studies in the Romance Languages and Literatures 58 (Chapel Hill, NC: University of North Carolina Press, 1965), pp. 115–40.

42. On the faculty represented by Imaginatif see Minnis, "Langland's Ymaginatif and late-medieval theories of the imagination," *Comparative Criticism* 3 (1981): 71–103, and more recently Ralph Hanna III, "Langland's Ymaginatif: Images and the Limits of Poetry," in *Images, Idolatry, and Iconoclasm in Late Medieval England: Textuality and the Visual Image*, ed. Jeremy Dimmick, James Simpson, and Nicolette Zeeman (Oxford: Oxford University Press, 2002), pp. 81–94.

43. Whatley's "*Piers Plowman*" provides a valuable survey of earlier work; see also Doxsee, "Trew Treuth"; Minnis, "Looking for a Sign," pp. 151–56; and Grady, "Rule," pp. 71–75. For the history of the "facere" ethic, see Obermann, "*Facientibus quod in se est Deus non denegat gratiam:* Robert Holcot, O.P., and the Beginning of Luther's Theology," *Harvard Theological Review* 55 (1962): 317–42. An interesting contemporary observation on this passage appears in the margin of MS Douce 104 (a C-text), where an annotator has written "nota of iiii follyng[gis]"—an additional piece of evidence favoring Whatley's contention (pp. 7–8) that the "Ac" at the beginning of line 287 indicates a move to a fourth kind of baptism. For the annotation see Kerby-Fulton and Despres, *Iconography and the Professional Reader*, p. 187.

44. Green, *Crisis of Truth*, pp. 3–9, 30, 31.

45. As Green (*Crisis of Truth*, pp. 23, 371) points out, there are places in the C-version where exactly the opposite is true, where Langland's revisions are designed to clear up the somewhat muddy significations of "truth" that could attach to the story of Trajan's salvation.

46. On the concept of God keeping his covenants see Green, *Crisis of Truth*, chap. 9, "Bargains with God."

47. This sense of "trowþe" as a creed issuing in ethical behavior is shared—and perhaps even borrowed—by Walter Hilton in the *Scale*, where he draws a much sharper line than Langland between God's truth and the truth achieved by non-Christians: "Sen þis is soþen þink me þat þese men gretly and greuously erre þat saien þat Iewes and Saraȝeins, bi keping of þeir own law, moun be mad saf þawȝ þei trowe not in Ihesu Crist als haly kirke trowes, in als mikel as þei wene þat þeir owne trowþis good and siker and suffisaunt to þair saluacioun, and in þat troub þei doo, as it semes, many gode dedes of riȝtwisnes, and perauenture if þei knewe þat cristen feiþ ware better þen þaires þei wold leue þeire own and take it þat þei þerfore schuld be saf. Nai, it is not inowȝ so. For Crist God and Man is boþ wei and ende, and he is mediatour bitwix God and man, and wiþouten hym may no soule be reconsilid ne come to þe blis of heuen. And þerfor þei þat trow not in hym

þat he is boþ God and man moun neuer be saf ne come to blis." Quoted in Watson, "Visions of Inclusion," p. 175 n. 8.

48. See Whatley, "*Piers Plowman*," pp. 2–5, where he argues that although it is usually glossed as "praise; commend; reward," *alowe* can have a specific theological meaning "denoting the action of divine grace that rewards an individual for his intention, his 'will,' rather that for his actual deed or conformity to a law or creed" (4).

49. Watson, "Visions," pp. 157–60.

50. I have abandoned Kane-Donaldson's "haluebreþeren" here in favor of the more widely attested "breþeren."

51. Watson, "Visions," p. 160.

52. Kane-Donaldson read "Grekes" for "Iewes," the more common reading here.

53. The words and the sentiment are Minnis's ("Looking," 162), though he goes on to argue that "the difference between the supposedly 'radical' thesis and the supposedly 'conservative' resolution may not be as great as some have supposed" (163). See Introduction n.11 for a similar argument about Aquinas.

54. Whatley, "Heathens and Saints: *St. Erkenwald* in Its Legendary Context," *Speculum* 61 (1986): 330–63, here p. 333. The poem, Whatley claims, is "an attempt to reinvest the legend with the hagiographical spirit of its earliest versions. Far from echoing the ideas of Langland's Trajan, the *St. Erkenwald* poet is utterly at odds with him and with everything he represents" (342). But cf. Ruth Morse, who in her edition of the poem (Cambridge: D.S. Brewer, 1975) argues that *Erkenwald* "tells the story of the righteous heathen in a way that implies certain liberal theological interpretations of Good Works and their claim upon God's mercy" (8).

All quotations are from Clifford Peterson's edition of the poem, *St. Erkenwald* (Philadelphia: University of Pennsylvania Press, 1977). For the date of *St. Erkenwald*, generally assumed to be later than the B-text of *Piers*, see note 64 later.

55. Whatley, "Heathens and Saints," pp. 333, 342. See also his "The Middle English *St. Erkenwald* and Its Liturgical Context," *Mediaevalia* 8 (1982): 278–306.

56. As Whatley acknowledges in "Heathens and Saints," p. 350.

57. *Paradiso* 19.77–78, 79–81, from Dante Alighieri, *The Divine Comedy: Paradiso*, trans. Charles S. Singleton (Princeton: Princeton University Press, 1975), 1:212–15.

58. Jim Rhodes, in a fine close reading of the last section of the poem, takes this notion of stage management yet further by suggesting that the miraculous salvation "cuts off dialogue prematurely," silencing the pagan "at a moment when he has leveled his most devastating remarks. . .at the very doctrine that purportedly has saved him, and at a point when he has exerted his most profound effect on the audience, having moved them all to tears." This observation may point us to the true nature of *Erkenwald*'s "conservatism," which could be said to lie not in its sacramental theology but in its willingness to use the sacrament in the service of an exceptional salvation; we know

from Chaucer's *Clerk's Tale*, for example, how effectively the spurious harmony of a happy ending can terminate a threat to the ideological status quo. See Rhodes, *Poetry Does Theology: Chaucer, Grosseteste, and the Pearl Poet* (Notre Dame, IN: Notre Dame University Press, 2001), pp. 157–58, and also 163–65.

59. Monika Otter draws similar conclusions about the role of history in the poem in " 'New Werke': *St. Erkenwald*, St. Albans, and the medieval sense of the past," *Journal of Medieval and Renaissance Studies* 24 (1994): 385–414.

60. They are, of course, looking in the wrong place, searching regal chronicles rather than legal records. T. McAlindon ("Hagiography into Art: A Study of *St. Erkenwald*," *Studies in Philology* 67 [1970]: 338) points out that most English legends involving uncorrupted bodies of saints did feature kings, Bede's St. Cuthbert being a notable exception. For a sophisticated account of the poem's meditation on death and the labor of memory, see D. Vance Smith, "Crypt and Decryption: *Erkenwald* Terminable and Interminable," *NML* 5 (2002): 59–85.

61. See for instance Whatley: ". . .human reason proves hopelessly inadequate to explain the enigma of the corpse and has to give way to the religious, sacramental approach of Erkenwald" ("Heathens and Saints," p. 338).

62. Probably Geoffrey of Monmouth's Belinus; see Morse, p. 72n.213.

63. Whatley, "Heathens and Saints," p. 337.

64. On this topic see Ruth Nisse, " 'A Coroun Ful Riche': The Rule of History in *St. Erkenwald*," *ELH* 65 (1998): 277–95.

65. As Ruth Morse notes in her edition of the poem, the saint appears in Bede, William of Malmesbury, Matthew Paris, the *Flores Historiarum*, and John of Tynemouth. There are also two Latin lives, one from the turn of the eleventh century (the *Vita sancti Erkenwaldi*) and one from the 1140s (the *Miracula sancti Erkenwaldi*), both of which have recently been edited by Whatley. For the prominence of Erkenwald's cult in the later Middle Ages, see Morse, pp. 13–15. The two proclamations by Bishop Braybrooke have long been the primary evidence in the debate over the poem's date, which Henry L. Savage placed ca. 1386 (pp. lxxv–lxxvi); Peterson argues for a somewhat later date in his edition (11–15). For an argument placing the poem in the context of political events ca. 1388–92, see my "*St. Erkenwald* and the Merciless Parliament," *SAC* 22 (2000): 179–211.

66. For the solution to Conscience's riddle in *Piers*, see Andrew Galloway, "The Rhetoric of Riddling in Late-Medieval England: The 'Oxford' Riddles, the *Secretum philosophorum*, and the Riddles in *Piers Plowman*," *Speculum* 70 (1995): 68–105, esp. pp. 86–90. Galloway's convincing argument that Conscience's "þe myddel of a Moone" refers to a Latin riddle whose solution is *cor*, that is, a conversion or transformation of the human heart, suggests that the pagan judge's line "Als ferforthe as my faithe confourmyd my hert" represents another allusion to *Piers Plowman*, and an extremely sophisticated one (maybe one too sophisticated to be likely).

67. The ornately decorated sepulchre also recalls the forum monument that presumably inspired Gregory's prayer for Trajan, with this difference: in the

original legend, Gregory correctly misreads the sculpture as celebrating Trajan's justice (if the hypothesis of Paris, Vickers, et al. is correct; see n. 7 above), while the Londoners *incorrectly* misread the tomb as referring to its inhabitant's regality.

68. Cf. Otter, " 'New Werke,' " pp. 412–14; and Smith, "Crypt and Decryption," p. 71.

69. See Smith, "Crypt and Decryption," pp. 83–85, for a version of this claim made along the mourning/melancholia axis.

Chapter 2   Mandeville's "Gret Meruaylle"

1. For a succinct account of the relevant facts about author and text, see M.C. Seymour, *Sir John Mandeville*, Authors of the Middle Ages 1 (Aldershot: Variorum, 1994). The best recent book on *Mandevilles' Travels* is Iain Higgins's *Writing East: The "Travels" of Sir John Mandeville* (Philadelphia: University of Pennsylvania Press, 1997), an encyclopedic (almost Mandevillean) account of the sources and textual and manuscript traditions of the *Travels* and an indispensable companion for anyone interested in the *Travels'* composition and reception.

2. These are the conclusions, in order, of J.W. Bennett, *The Rediscovery of Sir John Mandeville* (New York: Modern Language Association of America, 1954); Douglas R. Butturf, "Satire in Mandeville's Travels," *Annuale Medievale* 13 (1972): 155–64; Christiane Deluz, *Le livre de Jehan de Mandeville: Une "geographie" au XIVe siecle*, Textes, Etudes, Congres 8 (Louvain-la-Neuve: Institut d'études médiévales de l'Université catholique de Louvain, 1988); and Mary B. Campbell, *The Witness and the Other World: Exotic European Travel Writing 400–1600* (Ithaca, NY: Cornell University Press, 1988). In an important chapter on the *Travels* in *Empire of Magic: Medieval Romance and the Politics of Cultural Fantasy* (New York: Columbia University Press, 2003), Geraldine Heng also describes the work as a "travel romance," "a prime representative of that hybrid genre of inextricably conjoined, seamlessly indistinguishable fact and fantasy we have been calling romance" (240).

3. The most compelling versions of this claim are those of Donald Howard, "The World of Mandeville's Travels," *The Yearbook of English Studies* 1 (1971): 1–17, revised in Howard's *Writers and Pilgrims: Medieval Pilgrimage Narratives and Their Posterity* (Berkeley: University of California Press, 1980), chap. 3; and Christian K. Zacher, *Curiosity and Pilgrimage* (Baltimore: Johns Hopkins University Press, 1976), pp. 130–57.

4. On Mandeville's anti-Jewish and anti-Semitic tendencies, to which I return in my conclusion, see Higgins, *Writing East*, pp. 16–17 and 80–81; Stephen Greenblatt, *Marvelous Possessions: The Wonder of the New World* (Chicago: University of Chicago Press, 1991), pp. 50–51; Benjamin Braude, "*Mandeville's* Jews Among Others," in *Pilgrims and Travellers to the Holy Land*, ed. Bryan F. LeBeau and Menachem Mor, Studies in Jewish Civilization 7 (Omaha, NE: Creighton University Press, 1995), pp. 141–68; Kathleen Biddick, "The ABC of Ptolemy: Mapping the World with the Alphabet," in

Text and Territory: Geographical Imagination in the European Middle Ages, ed. Sylvia Tomasch and Sealy Gilles (Philadelphia: University of Pennsylvania Press, 1998), pp. 268–93; Scott D.Westrem,"Against Gog and Magog," in *Text and Territory*, pp. 54–75; and Linda Lomperis, "Medieval Travel Writing and the Question of Race," *JMEMS* 31 (2001): 162–63.

5. My inclusion of *Mandeville's Travels* in a book on English writing is based on the fact that at least six different English writers in the later Middle Ages tried to adopt the text as their own. See Higgins, *Writing East*, pp. 21–23.

Taking the view that I do of the text's structure does not preclude agreement with David Lawton's characterization of the *persona* of "Sir John": "The 'I' of *Mandeville*, and the text's embellishments, is for the most part a series of different first-person pronouns drawn directly from the different sources from which the account is compiled. . .There is no single subjectivity in the text but a series of subject positions, drawn from or responding to context, for readers to colonize in reading." Part of the text's genius is its collocation of subject positions sufficiently similar to encourage a relatively coherent process of readerly colonization—a process that continues in ongoing modern critical attempts to unearth the "real" Sir John Mandeville among extant medieval records. See David Lawton,"The Surveying Subject and the 'Whole World' of Belief:Three Case Studies," *NML* 4 (2001): 25.

6. See *Chronica Monasterii de Melsa*, ed. E.A. Bond, Rolls Series 43 (London: Longmans, Green, Reader, and Dyer, 1866–68), III.158–59. A better known chronicler,Thomas Walsingham, was also acquainted with the *Travels*, though he does not seem to have read it closely; though he includes "Johannes de Mandevile" in his list of famous sons of St. Albans, and even acknowledges that he composed his book in French, he also claims that Sir John took part "in many wars against the enemies of our faith" ("in multis bellis contra nostrae fidei adversarios"), a claim hardly sustained by the text. See *Annales Monasterii S. Albani*, ed. H.T. Riley, Rolls Series 28 (London: Longmans, Green, 1870–71), II.306. I owe both of these citations to David Wilmot Ruddy, "Scribes, Printers and Vernacular Authority: A Study in the Late-Medieval and Early-Modern Reception of *Mandeville's Travels*," Ph.D. Diss, University of Michigan, 1995.

7. All quotations from *Mandevilles' Travels* are taken from M.C. Seymour's edition of the Cotton text (Oxford: Oxford University Press, 1967), and are cited parenthetically by chapter and page number. My choice of Cotton as the representative English version here is partly practical and traditional—it is available in two relatively common critical editions, Seymour's and the EETS edition—and partly due to its relatively faithful and complete rendering of its French source.

8. William of Tripoli, "*Tractatus de statu Saracenorum*. . .," in *Kulturgeschichte der Kreuzzuge*, ed. Hans Prutz (1883): 573–98. William was a Dominican living in Acre in the second half of the thirteenth century. Of course neither William's tract nor the *Travels'* adaptation of it can be said to represent medieval Islam "as it was"; they offer Latin Christendom's conventionally positive rather than its conventionally derogatory point of view.

9. Caesarius of Heisterbach, *Dialogus Miraculorum*, ed. J. Strange (Cologne: H. Lempertz & comp., 1851), book IV chap. 15; see also *The Dialogue on Miracles*, ed. and trans. E. von E. Scott and C.C. Swinton Bland (London: G. Routledge & sons, ltd., 1929). On Mandeville's use of this episode, see Higgins, *Writing East*, p. 115; *Mandeville's Travels: Texts and Translations*, ed. Malcoln Letts, Hakluyt Society, 2nd ser. 101 (1953): 97n.2; and Victor Chauvin, "Le Prétendu Séjour de Mandeville en Égypte," *Wallonia* X (1902): 237–42.

10. Indeed, the conversation between Mandeville and the Sultan later took on a life of its own, as witnessed by the "Stanzaic Fragment" of Bodleian MS e Musaeo 160, an early sixteenth-century MS containing 313 lines about "[Ser Iohn Mandev]ille and Ser Marc of Venesse," including a report of "The Commonyng of Sir Iohan Mandeville and þe Gret Sowdon" (ll. 185–240). Seymour, who has edited the fragment ("Mandeville and Marco Polo: A Stanzaic Fragment," *AUMLA: Journal of the Australasian Universities Language and Literature Association* 21 [1964]: 39–52), sees a general but not certain correspondence between the verse and the Cotton version of the *Travels*. At the very least, the "Fragment" reveals a reader who found Mandeville interesting precisely because of his description of pagan kingdoms: the "Sowdan" and the "Grete Caan" occupy virtually every stanza.

11. *Dialogus Miraculorum*, iv. 187: "Ille dicere nolens quod verum fuit, respondit, 'Satis bene.' "

12. This theme first arises at prologue 2–3, and the author of the *Travels* returns to it again and again: "For God wole not that [the Holy Land] be longe in the hondes of traytoures ne of synneres, be thei Cristene or othere. And now haue the hethene men holden that lond in here hondes xl. yere and more, but thei schulle not holde it longe yif God wole" (10.55). And again: "But whan God allemyghty wole, right als the londes weren lost thorgh synne of Cristene men, so schulle thei ben wonnen ayen be Cristen men thorgh help of God" (10.58). And again: "For withouten ony drede, ne were cursednesse and synne of Cristene men, thei scholden ben lordes of alle the world" (28.188).

13. The Cotton version is actually very garbled here: "And also no straungere cometh before him but that he maketh hym sum promys and graunt of that the Soudan asketh resonably, be it so it be not ayenst his lawe. And so don othere prynces beyonden, for thei seyn that no man schalle come before no prynce but that the Soudan is bettre and schalle be more gladdere in departynge from his presence thanne he was at the comynge before hym" (VI.28). I have omitted for sense in the text. The French text (quoted by Hamelius [ii.43] and Seymour [260n.]) reads much more sensibly, as do the Egerton and Bodley versions (see Letts, i.28 and ii.434).

14. On the commonplace of Saracen plenitude in matters of wealth, luxury, and sex, see Jeffrey Jerome Cohen's reading of the contemporary Middle English romance *The Sultan of Babylon* in his "On Saracen Enjoyment: Some Fantasies of Race in Late Medieval France and England," *JMEMS* 31 (2001): 124–34.

15. "Et sic simplici sermone Dei, sine philosophicis argumentis sive militaribus armis, sicut oves simplices petunt baptismum Christi et transeunt in ovile Dei." *De Statu Saracenorum*, pp. 597–98.

16. The *Dialogus miraculorum* episode also contains a prophecy of the Christians' impending (but not immediate) reclamation of the Holy Land.

17. On the level of genre, the refusal of the princess is both a rejection of one of the commonplaces of Saracen romance and a generative romance moment; by declining to end his travels in this conventional way, Sir John leaves himself open to further adventure. Dorothy Metlitzki describes the importance of the marriage theme in literary treatments of Christian–Muslim relations (see *The Matter of Araby in Medieval England* [New Haven: Yale University Press, 1977], pp. 136–76); I argue below, however, that in *Mandeville's Travels* war and commerce—two sides of the same coin, as it were—bind the two cultures much more closely and permanently than love.

18. Heng, *Empire of Magic*, pp. 250, 242.

19. In Caesarius's account, the Saracen had been sent as a youth to the court of the King of Jerusalem "ut Gallicum discerem apud illum," while the King's son was sent to the Emir's father "ad discendum idioma Sarracencium." *Dialogus miraculorum*, iv.187.

20. Philippe de Mézières, *Le Songe du Vieil Pelerin* (2 vols., ed. G.W. Coopland [Cambridge: Cambridge University Press, 1969]), II. 405: "les espies par lesquelles on puet mieulx savoir l'estat de ses ennemis ce sont les marchans Lombars et estranges, qui pour la marchandie ont occasion en personne ou par leurs facteurs et par leurs lectres de mander de l'une parte et de l'autre; et par espicial par les marchans qui pour les pierres precieuses et joyaux ont entree et privee amitie avec les roys et princes."

21. Evidence of Western travellers fluent in Arabic does not seem to be extensive, although Norman Daniel does cite one anecdote from Geoffrey of Vinsauf's chronicle of Richard I's crusade in the 1190s that describes both the fluency and the Egyptian clothing of "Bernard, 'the king's spy.' " See Daniel, *The Arabs and Medieval Europe* (London: Longman, 1975), p. 199. For a more recent survey of the topic see Hussein M. Attiya, "Knowledge of Arabic in the Crusader States in the twelfth and thirteenth centuries," *Journal of Medieval History* 25 (1999): 203–13.

22. For this observation about the "commodification" of Sir John's skills I am indebted to Laura King. On the role of European and particularly English mercenaries in the Mediterranean, and the Muslim response to the doctrinal impediments involved, see my "Machomete and *Mandeville's Travels*," in *Medieval Christian Perceptions of Islam*, ed. John Tolan (New York: Garland Publishing, 1994), pp. 281–82 and p. 288 nn.30–32.

23. For another version of this gesture, without the empirical complications, see "Lord Cobham's Cliché" in chapter 4.

24. The first time is at 9.52, when he is explaining why the Saracens do not have vineyards.

25. Seymour, *Mandeville's Travels* (1967), p. 245 n.104/8.

26. He is of course well acquainted with the conventional moral myth of language differentiation; in his earlier account of the Sultanate he had carefully distinguished between the Babylon where the Sultan dwells (that is, Cairo) and "that gret Babyloyne where the dyuersitee of langages was first made for vengeance by the myracle of God" (6.28). Even there, however, he shows more interest in the metropolis than the miracle, describing at some length the dimensions of the city and describing Nimrod as the first idolator rather than a prideful overreacher.

27. On the text's rhythmic and pleasurable alternation between the exotic and the familiar, see Heng, *Empire of Magic*, pp. 254–58.

28. Higgins argues that "Hovering behind this portrait of the Khan. . . .is the dream of a universal Christendom" (168). See also Heng on the text's "optimistic hypothesis of a cosmopolitan Christianity beyond the border of Europe" (270).

29. As usual, this claim is based on a story that is partially true. During the Mongol expansion westward in the thirteenth century, Hulagu had captured Aleppo on January 25, 1260, but Mangu's death forced him to return to the east before he could advance on Damascus and Jerusalem. In his absence, the Mamluks were able to retake their lost territory. Thus the author of the *Travels* goes beyond (or simply wishfully misreads) his source, Hayton's *Fleurs des Histors D'Orient*, which reads thus: "Apres ce que Haloon [Hulagu] ot ordene ce que faisoit mestier entour la garde de la citei de Halape [Aleppo] e de Damas, e des autres terres entor, les queles il avoit conquises contre les Sarazins, *si come il entendoit* entrer au roiaume de Jerusalem por delivrer la Terre Sainte e rendu cele as Crestiens, vesci venir un messaige que lui conta come sa frere Mango Can estoit trespasse de cestire siecle, e come les barons le queroient por faire le empereor." See Hayton, *Fleurs des Histors D'Orient*, in *Recueil des Historiens des Croisades: Documentes Armeniens* ed. Éd. Dulaurier (Paris: Imprimerie Nationale, 1906), ii. 172; emphasis added. The author of the *Travels* also omits Hayton's account of what happens after Hulagu's departure: his deputy has a falling out with some Christians who have slain his nephew, and the subsequent rift in the alliance leads to the Saracens' recovery of their lost lands. See Morris Rossabi, *Khubilai Khan: His Life and Times* (Berkeley: University of California Press, 1988), pp. 54–55; and for Hayton's (or Hetoum's) career, see *A Lytell Cronycle: Richard Pynson's Translation (c. 1520) of La Fleur des histoires de la terre d'Orient (c. 1307)*, ed. Glenn Burger, Toronto Medieval Texts and Translations 6 (Toronto: University of Toronto Press, 1988), pp. x–xxiii.

30. On this passage, see Higgins, *Writing East*, pp. 168–69. The source of the prophecy is John de Plano Carpini, available via Vincent of Beauvais' *Speculum Historiale*. Hamelius cites the relevant passage—sans archers—at ii. 124; see also Letts, i. 174n.1, and Christopher Dawson, *Mission to Asia*, Medieval Academy Reprints 8 (Toronto: University of Toronto Press, 1980), pp. 25–26: "They have fought now for forty-two years and are due to rule for another eighteen years. After that, so they say, they are to be conquered by another nation, though they do not know which; this has been foretold them. Those who escape with their lives, they say, are to keep the same law

which is kept by those who defeat them in battle." John was not very interested in toleration; in fact, much of his *History of the Mongols* is given over to
warnings and advice to the West about a potential Mongol invasion, which
seemed possible in the mid-thirteenth century.

31. For discussions of fact, fiction, and belief in *Mandeville's Travels* see Heng,
Empire of Magic, pp. 288ff.; and for this passage in particular Campbell,
Witness, pp. 144–48.

32. On the shifting history of these assignments in both medieval thought and
modern scholarship, see Benjamin Braude, "The Sons of Noah and the
Construction of Ethnic and Geographical Identities in the Medieval and
Early Modern Periods," *The William and Mary Quarterly* 54 (1997): 103–42;
Braude discusses the *Travels* on pp. 115–20.

33. Hayton, ii. 150, calls this commandement "molt cruel"; note the lack of
editorializing in the *Travels*.

34. Compare, for instance, the citation of Exodus 23:15, "Nemo accedat in
conspectu meo vacuus," to explain the custom that "no straungere schalle
come before [the Great Chan] but yif he yeue hym sum manere thing. . ."
(25.176). A different but equally suggestive image, one that gives the Chan
himself divine associations, is this one: "And vnder the emperoures table sit-
ten iiii. clerkes that writen alle that the emperour seyth, be it good, be it
euylle. For alle that he seyth moste ben holden, for he may not chaungen his
woord ne revoke it" (23.157). The clerks seem to be equal parts chancery
scribes and evangelists.

35. Hayton's title for the chapter in which this miracle occurs leaves no doubt
as to the identity of "God inmortalle": "Comment Nostre Seigneur demonstra
a Canguis Can et a sa gent voye pour passer le mont de Belgian." Hayton,
Fleurs, p. 153.

36. For a summary of the Brahmans' appearances in medieval literature, see
Thomas Hahn, "The Indian Tradition in Western Medieval Intellectual
History," *Viator* 9 (1978): 213–34, and chapter 3 below. The Brahman
episode may be the most conventionally "Orientalist" moment in the
Travels, in Edward Said's terms; theirs is always imagined as a timeless, static
culture, as virtuous and nudist in Alexander's time as in the later Middle
Ages. See Edward Said, *Orientalism* (New York: Random House, 1994).

37. See chapter 3, and also Westrem, "Against Gog and Magog," whose work
supersedes A.R. Andersen, *Alexander's Gate, Gog and Magog, and the Inclosed
Nations* (Cambridge, MA: The Medieval Academy of America, 1932).
Zacher, *Curiosity and Pilgrimage*, p. 142, briefly notes the resemblance
between Mandeville and Alexander.

38. *Iter ad Paradisum*, ed. Julius Zacher (Konigsberg: T. Thiele, 1859); and Mario
Esposito, *Hermathena* XV (1909): 368–83. For a discussion of the sources of
this legend, which is of Hebrew origin, see George Cary, *The Medieval
Alexander*, ed. D.J.A. Ross (Cambridge: Cambridge University Press, 1956;
repr. 1967), pp. 18–21, and Paul Meyer, *Alexandre le Grand dans la littérature
française du moyen âge*, 2 vols. (Paris: F. Vieweg, 1886), ii. 47–51; for its
connection to the English Alexander texts see Mary Lascelles, "Alexander

and the Earthly Paradise in Mediaeval English Writings," *Medium AEvum* 5 (1936): 31–47, 69–104, 173–88, especially pp. 81–92.

39. On the issue of circumnavigation in the *Travels*, see Higgins, *Writing East*, pp. 132–39. On conventional medieval notions of geography, see G.H.T. Kimble, *Geography in the Middle Ages* (London: Methuen & Co., 1938) and Leonard Olschki, *Marco Polo's Precursors* (Baltimore: Johns Hopkins Press, 1942). E.G.R. Taylor, "The Cosmographical Ideas of Mandeville's Day," (in *Mandeville's Travels: Texts and Translations*, ed. Letts, I. li–lix), claims that Mandeville's position on circumnavigation argues for the author's university education, probably at Paris.

40. See for example David May, "Dating the English Translation of Mandeville's Travels: The Papal Interpolation," *NQ* n.s. 34 (1987): 175–78, and Heng, *Empire of Magic*, pp. 263–64. Higgins assumes, I think correctly, that the interpolation that appears in the Cotton text is an earlier form than that preserved in the Egerton and Defective versions.

41. *Mandeville's Travels* ed. Hamelius, ii. 146. For Odoric see Manuel Komroff, *Contemporaries of Marco Polo* (New York: Boni & Liveright, 1928), pp. 244–45.

42. Howard, "World," p. 16. See also Greenblatt, *Marvelous Possessions*, pp. 44–45, who finds the scene an equally remarkable "radical transvaluation of values."

43. This incident, too, may be derived from the Alexander romances, although Sir John's speculations are much more optimistic. Alexander and his armies reach "þe Occyan at þe erthes ende, & þare in an ilee he heres / A grete glauir & a glaam of Grekin tongis." Alexander orders some of his soldiers to disrobe and swim to the island, but they are all drowned by crabs, and the conqueror declines to pursue the matter further. See *The Wars of Alexander*, ed. Hoyt N. Duggan and Thorlac Turville-Petre, EETS s.s. 10 (Oxford: Oxford University Press, 1989), p. 174, ll.5629–32.

44. Higgins, *Writing East*, p. 256.

45. In fact this episode reworks material from chapter 16, where in his account of Trebizond Sir John tells the story of Athanasius: "In that cytee lyth Seynt Athanasie, that was bisshopp of Alisandre, that made the psalm *Quicumque vult*. This Athanasius was a gret doctour of dyuynytee, and because that he preched and spak so depely of dyuynytee and of the godhede, he was accused to the Pope of Rome that he was an heretyk; wherfore the Pope sente after hym and putte hym in presoun. And whils he was in presoun, he made that psalm and sente it to the Pope and seyde that yif he were an heretyk, than was that heresie, for that he seyde was his beleeue. And whan the Pope saugh it and had examyned it, that it was perfite and gode and verryly oure feyth and oure beleeue, he made hym to ben delyuered out of presoun and commanded that psalm to ben seyd euery day at pryme; and so he held Athanasie a gode man. But he wolde neuere go to his bisshopriche ayen because that thei accused him of heresye" (16.106–07).

46. Higgins concedes the resemblance in the late Latin Cosin MS, but to my mind that writer simply picks up on hints already present in the earlier versions of the interpolation.

Chapter 3   The Middle English Alexander

1. *Alexander B,* ll.81–82, 84, 91–110, in F.P. Magoun, *The Gests of King Alexander of Macedon* (Cambridge, 1929), pp. 174–75. Further references will appear in the text identified by line number.

2. Magoun, *Gests,* pp. 174–75. Both *Alexander and Dindimus* and *The Wars of Alexander* are descended from the *Nativitas et Victoria Alexandri Magni,* Archpresbyter Leo of Naples' tenth-century Latin translation of the Greek Alexander romance of Pseudo-Callisthenes (ca. 200 BC–AD 200). The *Nativitas,* which became known as the *Historia de Preliis* in its incunabular versions, also produced three interpolated (I) recensions: I^1, an eleventh-century reworking and expansion; I^2, of an uncertain date before the end of the twelfth century (the source of *Alexander and Dindimus* and *Alexander A,* another fourteenth-century Middle English alliterative poem dealing with the beginning of the legend); and I^3, from before 1150 (source of the *Wars of Alexander* and the fifteenth-century *Prose Alexander* in the Thornton Manuscript). Recensions I^2 and I^3 are independently descended from I^1, and according to D.J.A. Ross, "No version of the Alexander-romance has had a wider influence nor produced more vernacular progeny than this wretched little book" (*Alexander Historiatus: A Guide to Medieval Illustrated Alexander Literature,* Warburg Institute Surveys 1 [London: Warburg Institute, 1963], p. 47). There are also works in Latin, German, French, Italian, Swedish, Hebrew, Czech, Polish, Russian, and Magyar that derive from the various versions of the *Historia de Preliis,* suggesting a history of transmission that rivals that of *Mandeville's Travels.*

   Prior to the composition of the *Historia de Preliis,* Julius Valerius' *Res Gestae Alexandri Macedonis,* a fourth-century translation of Pseudo-Callisthenes, was the chief Latin source of the Alexander legends; its popularity was eclipsed in the later Middle Ages by both the *Historia* and a ninth-century *Epitome* of the *Res Gestae* itself. The Julius Valerius tradition is the source for most French Alexander poems, two fifteenth-century Scottish versions, and the Middle English *Kyng Alisaunder* (before 1330). The most complete explication of the complex lines of transmission of the Alexander story in the Middle Ages is still George Cary's *The Medieval Alexander.* Less daunting versions can be found in Magoun, *Gests,* pp. 15–62, and Ross, *Alexander Historiatus.* See also Duggan, "The Source of the Middle English *The Wars of Alexander,*" *Speculum* 51 (1976): 624–36.

3. For an account of the Alexander material in these two texts see G.H.V. Bunt, "Alexander and the Universal chronicle: Scholars and Translators," in *The Medieval Alexander Legend and Romance Epic: Essays in Honour of David J.A. Ross,* ed. Peter Noble, Lucie Polak, and Claire Isoz (Millwood, N.J.: Kraus International Publications, 1982), pp. 1–10, and Bunt, "The Story of Alexander the Great in the Middle English translations of Higden's *Polychronicon,*" in *Vincent of Beauvais and Alexander the Great: Studies on the Speculum Maius and Its Translations into Medieval Vernaculars,* ed. W.J. Aerts, E.R. Smits and J.B. Voorbij, Medievalia Groningana fasc. 8 (Groningen: E. Forsten, 1986),

pp. 127–40. Bunt argues that Vincent was an important source for Higden's account of Alexander, though probably not his chief source, adducing as part of his evidence their divergent treatments of the Alexander–Dindimus episode.

4. For a concise account of this binary and its origins, see Elisa Narin van Court, "Socially Marginal, Culturally Central: Representing Jews in Late Medieval English Literature," *Exemplaria* 12 (2000): 293–326, esp. pp. 300–03.

5. *The Monk's Tale*, 2631–33, in *The Riverside Chaucer*, ed. Larry D. Benson et al., 3rd edn. (Boston: Houghton Mifflin, 1988).

6. Leo continues, "For the prelates, who are eke guides, by reading and meditating how the aforementioned pagans, idolaters though they were, bore themselves above reproach in all they did, may thus sharpen their minds by that example, with the resolve that they be known as leal members of Christ, and greatly outdo those others in chastity, righteousness, and piety." ("Certamina vel victorias excellentium virorum infidelium ante adventum Christi, quamvis exstitissent pagani, bonum et utile est omnibus Christianis ad audiendum et intelligendum tam praelatis quam subditis, vidilicet saecularibus et spiritualibus viris, quia cunctos ad meliorem provocat actionem. Nam prelati, id est rectores, legendo et considerando, quemadmodum praedicti pagani idolis servientes agebant se caste et fideliter atque in omnibus se inreprehensibiliter ostendebant, per eorum exempla bonorum operum ita acuant mentes suas, eo quod fideles et membra Cristi esse videntur, ut multo magis meliores se illis demonstrent in castitate et iusticia atque pietate.") The translation is Margaret Schlauch's, from *Medieval Narrative: A Book of Translations* (New York: Prentice-Hall, Inc. 1928), p. 285; for the Latin text, see *Die Historia de Preliis Alexandri Magni-Synoptische Edition der Rezensionen des Leo Archipresbyter und die interpolierten Fassungen J¹, J², J³ (Buch I & II)*, ed. Hermann-Josef Bergmeister, *Beitrage zur Klassischen Philologie*, Heft 65 (Meisenheim am Glan: A. Hain, 1975), p. 2a. The sentiments are as old as Augustine's *City of God*; compare his remarks on Roman virtue in book 5, chaps. 16–18, in *City of God*, trans. Henry Bettenson (New York: Penguin, 1980), pp. 205–12.

7. "Religiosis insuper et in claustris residentibus non erit hec compilacio minus utilis, qui ex processu delectacionem non modicam generit, et fortasse talium personarum tollet accidiam et tedium relevabit. Et cum sit generativa leticie lectio hujus hystorie, occasiones vagandi inutiliter aufert et efficaciter perimet et extinguet." Quoted from British Museum MS Douce 299, the unique copy of Walsingham's text, by Paul Meyer, *Alexandre le Grand dans la Litterature Francaise du Moyen Age* (Paris: F. Viewig, 1886), II: 65. On Walsingham's authorship of the later compilation and its date, see V.H. Galbraith, ed., *The St. Albans Chronicle 1406–1420* (Oxford: Clarendon Press, 1937), pp. xli, xliv–xlv, and James G. Clark, "Thomas Walsingham Reconsidered: Books and Learning at Late-Medieval St. Albans," *Speculum* 77 (2002): 832–60. For another version of the argument that reading and writing history, pagan and Christian, can relive the tediousness of the monastic life, see the *Eulogium Historiarum sive Temporis*, ed. Frank Scott Haydon, 3 vols., Rolls Series 9 (London: Longman, Brown, Green, Longmans and Roberts, 1858–63), I.3.

8. Frank Grady, "The Literary and Political Recuperation of Pagan Virtue in the English Middle Ages," Ph.D. diss. U.C. Berkeley, 1991, pp. 97–100, surveys the surviving records. In addition to the sources cited there, see also David N. Bell, ed., *The Libraries of the Cistercians, Gilbertines, and Premonstratensians*, CBMLC 3 (London: The British Library, 1992), p. 103 (for the Cistercian abbey of Rievaulx, Yorkshire), and R. Sharpe, J.P. Carley, R.M. Thomson, and A.G. Watson, *English Benedictine Libraries: The Shorter Catalogues*, CBMLC 4 (London: The British Library, 1995), pp. 12 (for Bardney, Lincolnshire) and 255–56 (for Holme St. Benets, Norfolk).

9. The manuscript is described by M.R. James in *A Descriptive Catalogue of the Manuscripts in the Library of Corpus Christi College, Cambridge* (Cambridge, 1909–12), II. 364–72. Given the contents of the volume and the references to Durham, James suggests that the book belonged to a "notarial personage of that diocese," (364) a supposition seconded by C.R. Cheney in his study of the MS, "Law and Letters in Fourteenth-Century Durham: A Study of Corpus Christi College, Cambridge, MS. 450," in his *The English Church and its Laws, 12th–14th Centuries* (London: Variorum, 1982), pp. 60–86. The letters appear on pp. 279–80 of the MS; the relevant chapters of Vincent are IV. 66 (to "in pace dimisit"—a passage not drawn from the *Collatio*), 67 (from "de nobis dudum" to the end), 68 (from "pestilentiam nos Bragmani" to the end), 69 (from the beginning to "minime frequentamus," omitting "Qui dum putant. . .perdidisse") and 70 (all of Dindimus's response, omitting "Deos autem. . .negamus admittere"). See *Bibliotheca Mundi seu Speculi Maioris Vincenti Bvrgundi Praesvlis Bellovacensis. . .*(Douai, 1624; repr. Graz: Akademische Druck-u Verlaganstalt, 1964–65), IV. 135–36.

10. Sharpe et al., *English Benedictine Libraries*, p. 395.

11. For the life and work of John of London, a mathematician who may have been this donor, see *DNB* x.885. M.R. James, *The Ancient Libraries of Canterbury and Dover* (Cambridge: Cambridge University Press, 1903), pp. lxxiv–lxxvii, suggests that there were two Johns, both mathematicians, both known to Bacon, and that the younger of the two (d. c. 1331?) was the man in question. He cites John as the donor of the *Gesta* (#916) on p. 295. This is British Library MS Additional 48,178 (Yelverton MS Appendix I); see *The British Library Catalogue of the Additions to the Manuscripts 1951–55* (London: The British Library, 1982), i.141, and *The British Library Catalogue of the Additions to the Manuscripts: The Yelverton Manuscripts* (London: The British Library, 1994), i.373–75, which dates the MS after 1316. James (pp. 375, 521) identifies the catalogue's #1544 (Liber de alkemia) with Glasgow, Hunterian MS 253 (U.4.11), though he does not associate the MS with John; however, a note in a fourteenth- or fifteenth-century hand on f.3ʳ of the MS ascribes it to John of London, so it ought to be counted as one of his donations. See John Young and P. Henderson Aitken, *A Catalogue of the Manuscripts in the Library of the Hunterian Museum in the University of Glasgow* (Glasgow: J. Maclehose and Sons, 1908), p. 206.

12. K.W. Humphreys, ed., *The Friars' Libraries*, CBMLC 1 (London: The British Library, 1990), pp. 120 (#486) and 114 (#463f). On Erghome's library see

Humphreys, "The Library of John Erghome and personal libraries of the fourteenth century in England," *Proceedings of the Leeds Philosophical and Literary Society*, Literary and Historical Section, 18 (1982): 106–23. For Erghome's commentary, see most recently A.G. Rigg, "John of Bridlington's Prophecy: A New Look," *Speculum* 63 (1988): 596–613.

13. For Hereford, see W.A. Hulton, ed. *Documents Relating to the Priory of Penwortham, and other Possessions in Lancashire of the Abbey of Evesham* (Manchester: Chetham Society, 1853), p. 95; Sharpe et al., *English Benedictine Libraries*, p. 144. This Hereford should not be confused with the sometime Lollard Nicholas Hereford, who died sometime after 1417. Wivill's book is Oxford, Worcester College Library MS 285; see R.W. Hunt, "A Manuscript Belonging to Robert Wivill, Bishop of Salisbury," *Bodleian Library Record* 7 (1962–67): 23–27. For Stafford, see *DNB* xviii.862–63. The erased inscription identifying him as donor is apparently readable under ultraviolet light, according to the on-line catalogue at Glasgow University Library (on 4/24/00): http://Special.lib.gla.ac.uk/manuscripts/detaild.cfm?DID=32625. For the Cotton MS, see H.L.D. Ward, *Catalogue of Romances in the Department of Manuscripts in the British Museum*, 3 vols. (London: Trustees of the British Museum, 1883–1910), i.114.

14. Madeleine Blaess, "L'Abbaye de Bordesley et les Livres de Guy de Beauchamp," *Romania* LXXVIII (1957): 511–18.

15. Viscount Dillon and W. St. John Hope, "Inventory of the goods and chattels belonging to Thomas, Duke of Gloucester. . .," *The Archaeological Journal* 54 (1897): 275–308.

16. Kathleen L. Scott, *Later Gothic Manuscripts, 1390–1490* (London: H. Miller, 1996), ii.73.

17. Ward, *Catalogue of Romances*, i.129; see also George Warner and Julius P. Gilson, *Catalogue of the Western Manuscripts in the Old Royal and King's Collections* (London: Trustees of the British Museum, 1921), iii.177–79. The manuscript survives as BL MS Royal 15.E.vi.

18. Walworth's interest is revealed in the note published by J.M. Manly in the *Times Literary Supplement* no. 1338 (Thursday, September 22, 1927), p. 647: "Students of the Alexander romances may be interested in the following entry from the Plea and Memoranda Rolls of the City of London (25m5d). I quote from the manuscript calendar, p. 155: 'Friday [27 Feb] after the feast of St. Mathias, 5 Richard II, came to the court held before the sheriffs of London, viz. on the fourth day, William Waleworth suing John Salman, a burgess of Bruges, in a plea of debt upon demand of 100L, for which his debtor has been attached by a foreign attachment and has made four defaults. The plaintiff demands his attachment on the customary terms and the articles are delivered to him, viz. a book of Romance of King Alexander in verse [*rimiatus*] and curiously illuminated, value 10L and a dosser of assar work, presenting the coronation of King Alexander, 9 yds. long, 3 yds. wide, value 6L.' Perhaps it may be possible to ascertain whether this copy of the romance is among those still preserved and known to us." M.R. James briefly considers the chances that the book referred to is actually MS Bodley 264, but

discounts the possibility; see James, *The Romance of Alexander: A Collotype Facsimile of MS Bodley 264* (Oxford: Clarendon, 1933), p. 5.

For Robert Thornton see George R. Keiser's two essays, "Lincoln Cathedral MS.91: Life and Milieu of the Scribe, *Studies in Bibliography* 32 (1979): 158–79, and "More Light on the Life and Milieu of Robert Thornton," *Studies in Bibliography* 36 (1983): 111–19. The text of the *Prose Alexander* has been edited three times: without apparatus by J.S. Westlake, EETS o.s. 143 (London: Keegan Paul, Trench, Trübner and Co., 1913); by Marjorie Neeson, "*The Prose Alexander:* A Critical Edition" Ph.D. Diss., UCLA, 1971; and by Julie Chappell, "The Prose 'Alexander' of Robert Thornton: The Middle English Text with a Modern English Translation," Ph.D. Diss., University of Washington, 1982. This last has also been published in book form (New York, 1992).

19. Thomas too had named one of his sons—one of John's younger brothers—Alexander. His library included at least two texts on the Trojan war, a volume of Petrarch, and a "librum parvum vocatum Scropp." This seems to have been a pocket book owned by Richard Scrope, the Archbishop of York executed by Henry IV in 1405, and treated as something of a relic by the Dautrees; in his will John leaves to his "spiritual father" William Langton, rector of St. Michael's Ousebridge, York, a book that "Beatus Ricardus le Scrop habuit et gerebat in sinu suo tempore suae decollacionis." William was to have the use of the book for the duration of his life, at the end of which it was to be returned to the site of Scrope's tomb, "ibidem pro remanere." They sound like an interesting family. For Thomas's will (1437) and John's (1459), see *Testamenta Eboracensa*, ed. J. Raine, Surtees Society 30 (1855), pp. 59–61, 230–34.

20. This text has been edited twice: see Thomas Hahn, "The Middle English *Letter of Alexander to Aristotle*: Introduction, Text, Sources, and Commentary," *Medieval Studies* XLI (1979): 106–45, and Vincent Dimarco and Leslie Perelman, eds., "The Middle English Letter of Alexander to Aristotle," *Costerus* 13 (1978). On the copyist of this manuscript, who according to A.I. Doyle was active during the reign of Edward IV, see Doyle's "An Unrecognized Piece of *Piers the Ploughman's Creed* and Other Work by Its Scribe," *Speculum* 34 (1959): 428–36, and Linne R. Mooney, "A New Manuscript by the Hammond Scribe Discovered by Jeremy Griffiths," in *The English Medieval Book: Studies in Memory of Jeremy Griffiths* ed. A.S.G. Edwards, Vincent Gillespie, and Ralph Hanna III (London: The British Library, 2000), pp. 113–23.

If we include scribes we could also add to this list Richard Frampton, the early fifteenth-century London scribe who copied the *Historia de Preliis* at least twice, in Cambridge University Library MS Mm.V.14 and Hunterian 84. For Frampton see A.I. Doyle and M.B. Parkes, "The Production of Copies of the *Canterbury Tales* and the *Confessio Amantis* in the Early Fifteenth Century," in *Medieval Scribes, Manuscripts and Libraries: Essays Presented to N.R. Ker,* ed. Parkes and Andrew G. Watson (London: Scolar Press, 1978), pp. 163–210, esp. pp. 192–95 and n.65.

21. Corpus Christi College, Cambridge MS 219 and Gonville and Caius College MS 154 (the latter from Bury; see Andrew Watson, *Medieval Libraries of Great Britain. . . .:A Supplement to the Second Edition* [London: Offices of the Royal Historical Society, 1987], p. 60).The most complete account of the *Compilation*, which has never been published in full, can be found in Meyer, *Alexander le Grand* II: 52–68; see also Cary, *Medieval Alexander*, pp. 68–69; and Magoun, *Gests*, pp. 244–54, where he prints a small portion of the text.The manuscripts are described in James's catalogue for Corpus Christi and his *A Descriptive Catalogue of the Manuscripts in the Library of Gonville and Caius College, Cambridge*, 2 vols. (Cambridge: Cambridge University Press, 1907–08).

22. D.J.A. Ross, "*Parva Recapitulatio*: An English Collection of Texts Relating to Alexander the Great," *Classica et Medievalia* 33 (1981–82): 191–203.

23. Cambridge University Library MS Ll.I.15, ff. 136^v–137.This addition is not, as the catalogue claims, "a continuation of the letter in the third person" (*A Catalogue of the Manuscripts Preserved in the Library of the University of Cambridge* [1854; repr. Munich: Kraus Reprint, 1980], iv.11), nor, as a marginal note in the manuscript suggests, a passage borrowed from Alexander Neckham's *De Naturis Rerum* (f. 136^v), but a bit of reconstructive surgery in the same vein as the inclusion of *Alexander and Dindimus* in Bodley 264.

To this list of "corrected" Alexander MSS we can also add BL MS Arundel 123, which contains a fourteenth-century copy of the I^1 recension of the *Historia de preliis*. At the point where Alexander encloses Gog and Magog—a story that only appears in certain I^1 MSS—the scribe also cites the version of the story told in Peter Comestor's *Historia Scholastica*, in order to stress Alexander's dependence on God's intervention. See G.H.V. Bunt, "The Art of a Medieval Translator: The Thornton Prose Life of Alexander," *Neophilologus* 76 (1992): 154.

24. Meyer, *Alexandre le Grand*, II.63–68, discusses the Douce MS; the final exchange with Dindimus begins on f.91^v, near the marginal note "Ranulphus." Another manuscript now at Cambridge, Gonville and Caius College MS 230/116, demonstrates that interest in Alexander continued through the fifteenth century at St. Albans; in addition to a number of the-ological tracts and sermons from the monastery, the manuscript preserves more than a dozen letter formulae, an English version of the *Carta humane redempcionis* (Christ's last will and testament, addressed to humankind), a tract entitled *Composicio Cartarum et aliarum euidentiarum*, and a *Littera increpatoria darii ad Alexandrum*, together with Alexander's response.The letter, in which Darius rebukes Alexander, calling him "famulus," slave, and sending him children's toys, is a well-known part of the correspondence between these two, the first step toward Darius's defeat by Alexander's forces; Alexander's reply interprets the toys (a ball, a string for a spinning top, and two small purses) figurally as signs of his impending conquest.The letter's preservation in a collection that includes several *inuectiones* of John Whethamstede (abbot from 1420 to 1440 and again from 1452 to 1465) argues for a strong sense of the relevance of the Alexander letters to the other missives of the same

heated tone and general purpose, and thus a continuing interest in classical writing at St. Albans. On Walsingham's classicism see Clark, "Thomas Walsingham Reconsidered." The "littera increpatoria" begins on f. 169; I have not yet been able to discover the specific source, although versions of the letter appear in the *Historia de Preliis* and in Julius Valerius. For a description of the manuscript and its contents, see M.R. James, *A Catalogue of. . .Gonville and Caius College,* I. 268–76. For a brief account of Whethamstede's career see David Knowles, *The Religious Orders in England II: The End of the Middle Ages* (Cambridge: Cambridge University Press, 1955), II.193–97, and for a survey of his literary output, R. Weiss, *Humanism in England During the Fifteenth Century* (Oxford: Blackwell, 1957), pp. 30–38.

25. Anne Middleton, "The Audience and Public of 'Piers Plowman,' " in *Middle English Alliterative Poetry and Its Literary Background,* ed. David Lawton (Cambridge: D.S. Brewer, 1982), pp. 104, 109. She continues, "These narratives comprise a kind of mythography of rule, a legendary for 'possessioners', lay and ecclesiastical. Whether at the level of catechitical dialogue (between ruler and philosopher, prophet and unjust judge, Saracen or pagan king and apostle) or war between the traditional culture-bearing power of the west and the opposing culture it will in the long process of time supplant (Greece and Troy, Rome and Jerusalem), these works present in a compendious historical mirror man's confrontation with spiritual dangers, and the rationale of large communal enterprises. They do so for the benefit, and from the viewpoint, of those who are situated to reflect on these enterprises and their consequences: the nobles, knights, burgesses and clerics who advised, judges, and acted in them by virtue of 'possessioun'—their responsibilities and powers devolve from what they hold and of whom" (109–10).

26. Middleton, "The Audience," p. 110.

27. Cary, *Medieval Alexander,* pp. 18, 125–30.

28. "Ay moȝt he lefe" is the English poet's hyperbolic rendering of the Latin's "vivat"; the addition is more than a little ironic, given that Iaudas has just been informed that Alexander will ultimately be "diȝt to þe deth" by God. Thornton translates simply "Lyff, lyffe" (f.6v).

29. Cf. the dream and explanation at ll. 1468–83, predicting the fall of Tyre to Alexander's siege. Mandeville inherits the Chan's dream from Hayton (see chapter 1); it might prove interesting to speculate about the influence on Hayton of the Alexander material, which certainly circulated as vigorously in the eastern Mediterranean as it did in England. The dream sequence is clearly a recurring motif in Western accounts of the foundation of pagan empires (and, in the Bible, of their disintegration), and it seems to have a double import: it demonstrates quite conventionally that God is firmly at the helm of world events, but it also suggests to the readers of *Wars* or Mandeville that God is just as interested as they are in the history of long ago times and far away places, and for some of the same reasons.

30. On theologians' efforts to disparage this gesture, see Cary, *Medieval Alexander,* pp. 128–29 and 294–95.

31. Gower retells this widely known story in book 3 of the *Confessio amantis*; see chapter 4.

32. For the suggestion that Alexander's generosity as recreated by the Jewish chroniclers is based on an actual "exemption fiscale" of Caesar's, see M. Simon, "Alexandre le Grand, juif et chretien," *Revue d'Histoire et Philosophie Religieuses* XXI (1941): 179. Simon makes an even stronger case than I do for the "Hebraicization" or "Christianization" of Alexander in this episode, claiming that "En se mettant sous son patronage, Israel annexe du même coup a sa foi: protecteur des Juifs, il est aussi l'instrument, elu et conscient, de Dieu" (180).

33. For the standard medieval interpretation of Daniel's prophecies as applying to Alexander see *Wars* 216 n.1779, and Cary, *Medieval Alexander*, pp. 119–21, 292–93. Early in the Jerusalem episode, Iaudas learns in a dream that there is a divine plan for Alexander:"For he mon ride þus & regne ouire all þe ronde werde, / Be lord [of] ilka lede into his laste days, / And þen [b]e diʒt to þe deth of driʒtin[e]s ire" (*Wars*, 1625–27).

34. Simon, "Alexandre," p. 190. Christine Chism makes an analogous suggestion at the level of narrative, arguing that the Jerusalem episode in *Wars* reveals a "manipulable" Alexander and that the Jews "are particularly perceptive in playing to Alexander's more grandiose self-conceptions" (*Alliterative Revivals* [Philadelphia: University of Pennsylvania Press, 2002], p. 144).

35. Had there been no such story, it would have been necessary to invent one—and in fact that is virtually what happened in the case of the short text D.J.A. Ross has entitled *Parva Recapitulatio de Eodem Alexandro et de Suis*. This piece, found in six English manuscripts, recounts some of the circumstances surrounding Alexander's birth, describes the battles of his successors, and in between and at greatest length retells the Jerusalem episode. It is always found in conjunction with the ninth-century *Epitome* of Julius Valerius's history of Alexander that otherwise lacks the Jerusalem material. The earliest version of the *Parva Recapitulatio*, which was probably translated directly from the Latin version of Josephus, appears in a late eleventh-century manuscript; in Ross's words, it "appears therefore to have been composed, probably in England, to correct the legendary texts in the collection [the *Epitome* was often collected with the *Epistola Alexandri ad Aristotelem* and the *Collatio cum Dindimo*] on the subject of Alexander's birth and parentage; and to supplement them with an account of the very popular visit to Jerusalem and a brief history of the struggles of Alexander's successors." See Ross, "*Parva Recapitulatio*," p. 194. Dimarco and Perelman claim that there is a Middle English translation of the *Parva Recapitulatio* in Worcester Cathedral MS F.172, alongside the Middle English version of the *Epistola Alexandri ad Aristotelem*. See "The Middle English Letter," pp. 5–6. The appearance of this text around the time of the First Crusade suggests that a favorable conception of Alexander included the recognition of his status as a quasi-crusader, liberating a grateful Jerusalem from the evil Persian empire. (For the possibility that Alexander's entry into Jerusalem as described by Vincent of Beauvais might have been an inspiration to the crusading Louis

IX [to whom the *Speculum Historiale* was dedicated], see J.B.Voorbij, "The *Speculum Historiale*: some aspects of its genesis and manuscript tradition," in *Vincent of Beauvais and Alexander the Great,* ed. Aerts et al., p. 31.) And that the *Parva Recapitulatio* continued to be copied through the fifteenth century testifies to the enduring popularity of this visit to Jerusalem, and to the extent of the peculiarly English syncretism visible in manuscripts like CUL Ll.I.15 and Bodley 264. Its popularity is also attested by the account of Alexander's career given by John Capgrave in his *Abbreuiacion of Cronicles,* which dates from the 1460s (ed. Peter J. Lucas, EETS o.s. 285 [Oxford: Oxford University Press, 1983]). Capgrave has two entries for Alexander, one which recounts his conversations with Diogenes and Demosthenes, and the second dealing with everything else:

> Anno 4879. Here deyed Grete Alisaundre, þat regned xii ȝere, sex ȝere with Darie, and sex ȝere aftir his deth. (And here leue we þe maner of countyng vsed befor, where we sette euyr the regner in his last ȝere; fro þis tyme forward we wil set hem in her first ȝere.) In þe sexte ȝere of Darie Alisaundre rejoysed þe kyngdam of Babilon, þat was þan, as we seid before deuolute to þe kyngdam of Perse, and now to þe kygdam of Macedonie. Thus was Alisaundre brout to þat empire, and sette mech good reule in euery lond. He visited þe Temple in Jerusalem and relesed hem of her tribute euery vii ȝere. He deyed in Babilonie, poisoned with venim. (42)

Lucas identifies Isidore of Seville as Capgrave's major source for pre-Roman history; his *Chronica Maiora* describes how Alexander "Hierusolymam capit atque templum ingressus deo hostian immolavit" but makes no mention of the suspension of tribute (See Isidore of Seville, *Chronica Maiora Isidori Iunioris,* ed. T. Mommsen, *Monumenta Germaniae Historica, Auctores Antiqquissimorum* XI, 2 vols. [Berlin: Weidman, 1884], ii.449). Given the English *Parva Recapitulatio* tradition, and Capgrave's positive account of Alexander's career, we might see his reference to the incident as an act of memory, the inclusion of a generally-known and generally-accepted fact about Alexander. When English readers and writers of Alexander's life determined what was integral and indispensable to the story, they took steps to supply it when necessary, whether it told of Alexander's celestial exploits, his philosophical disputations, or his intercourse with the holy sites and texts of the Judaeo-Christian tradition.

36. See Seymour, *Mandeville's Travels* (1967), chap. 29; *Wars,* ll.5609–28; *Prose Alexander,* ed. Westlake, pp. 104–05.

37. Of the twenty surviving manuscripts, which date from the mid-fourteenth through the end of the fifteenth century, six are fragmentary or incomplete, two represent a textual tradition that does not contain the two exempla (the *Expanded Northern Homily Cycle* in a Northern dialect), twelve contain the tale of Trajan's salvation, and ten of those pair Trajan with Alexander: Bodley 6923, CUL Dd.I.1, CUL Gg.V.31, Minnesota Z 822 N 81, Huntington HM 129, Lambeth 260, the Bute MS, BL MS Additional 38010, the Vernon MS, and

the Simeon MS. For an account of the MSS, see Saara Nevanlinna, ed., *The Northern Homily Cycle: The Expanded Version in MSS Harley 4196 and Cotton Tiberius E. VII*, Memoires de la Société Néophilologique de Helsinki 38 (1972). For the genesis and circulation of the cycle, see *The Idea of the Vernacular: An Anthology of Middle English Literary Theory, 1280–1520*, ed. Jocelyn Wogan-Browne, Nicholas Watson, Andrew Taylor, and Ruth Evans (University Park, PA: Pennsylvania State University Press, 1999), pp. 125–26, and Thomas J. Heffernan, "Orthodoxies Redux: The *Northern Homily Cycle* in the Vernon Manuscript and Its Textual Affiliations," in *Studies in the Vernon Manuscript*, ed. Derek Pearsall (Cambridge: D.S. Brewer, 1990), pp. 75–87. The one other Middle English text I know of in which these Alexander and Trajan stories appear together is James Yonge's *The Gouernance of Prynces*, a 1422 translation of the *Secretum secretorum*; I discuss this text in the conclusion.

38. Nevanlinna, *Northern Homily Cycle*, prologue, ll. 39–44, in *Idea of the Vernacular*, p. 127. The editors provide a translation: the cleric has "knowledge of God's word, /(For he has within him God's store / Of wisdom and of spiritual teaching / So that he ought not to spare a single bit of it / But reveal it to laymen / And make known to them the way to heaven)."

39. *The Vernon Manuscript: A Facsimile of Bodleian Library, Oxford, MS Eng. Poet. a. 1.*, intr. A.I. Doyle (Cambridge: D.S. Brewer, 1987), p. 11. The A-version of *Piers*, of course, lacks the appearance of Trajan, which is in B.XI.

40. I quote from the version of the sermon in the Vernon MS, and have transcribed passages directly from the manuscript facsimile prepared by Doyle, marking expanded abbreviations and supplying minimum punctuation; passages are identified by folio and column. This version of the sermon has not been printed in full, although the *exempla* are printed by Carl Horstmann, "Die Evangelien-Geschichten der Homiliensammlung des MS. Vernon," *Archiv für das Studium der neueren Sprachen und Literaturen* 57 (1877): 241–316.

41. The Vernon account of Trajan's rescue appears on f. 211r; a printed version can be found in Horstmann, "Evangelien-Geschichten," p. 306. The NHC version shares several important features with Langland's version: the Pope definitely prays for Trajan's soul; there is no question of revivification; Trajan's regal and courtly virtues are underscored, while the story of his kindness to the widow is omitted. The major difference, of course, is one of emphasis; the exemplum is offered as proof of Gregory's "treuþe," not Trajan's, and the Pope's part in their transaction is correspondingly stressed. This shift of narrative focus back to Gregory returns to the basic hagiographical design of the story. But it also reveals a profound truth about Trajan's place in the imagination of the poet-homilist. Although Gregory is chided briefly for making his request "unskilfulli," there is no suggestion that he is punished, or that his request presented any kind of doctrinal difficulty. For the versifier of this *exemplum*, the story of Trajan's exceptional salvation is an utterly unproblematic and entirely literal illustration of the power of "treuþe" to get one's prayers answered.

42. This story is based on a confusion of two other legends, both of Jewish origin; a condensed account of their origin and transmission can be found in Cary, *Medieval Alexander*, pp. 130–32, 295–97, and a fuller treatment in Andrew Runni Anderson, *Alexander's Gate, Gog and Magog, and the Inclosed Nations* (Cambridge, MA: Medieval Academy of America, 1932). A necessary supplement to the latter is Scott D. Westrem's "Against Gog and Magog," in *Text and Territory*, ed. Sylvia Tomasch and Sealy Gilles (Philadelphia: University of Pennsylvania Press, 1998), pp. 54–75. Horstmann, "Evangelien-Geschichten," pp. 306–07, prints the NHC version.

Josephus describes how Alexander imprisoned the Scythians behind iron gates, identifying the prisoners with the biblical giants—or peoples—Gog and Magog. Picked up in the Revelationes of Pseudo-Methodius, this version of the story passed into the I² and I³ versions of the *Historia de Preliis* and thence into the *Wars of Alexander*; see Alfons Hilka, *Der altfranzösiche prosa-Alexanderroman nach der Berliner bilderhandschrift. . .*(Halle a S.: M. Niemeyer, 1920), p. xxxi, and Karl Steffens, ed., *Die Historia de preliis Alexandri Magni: Rezension J3, Beiträge zur klassischen Philologie*, Heft 73 (Meisenheim am Glan: A. Hain, 1975), p. 174. A second story, often confused or fused with the first, involves Alexander's enclosure of the Ten Tribes and their apocalyptic fate; this is the version contained in the Vernon sermon. Delivered to the Middle Ages via the *Vitae Prophetarum* of Pseudo-Epiphanius, it appears in the *Historia Scholastica*, whence it passed to later chroniclers: Matthew Paris, Vincent of Beauvais (who treats the story with some scepticism), Higden and others. See Cary, *Medieval Alexander*, p. 296.

43. This rhetorical tag was falsely ascribed to Josephus, where it does not appear, and passes through the *Historica Scholastica* to Vincent of Beauvais: "Hic addit Josephus dicens, Deus quid facturus est pro fidelibus suis, si tantum fecit pro infideli?" See Vincent, *Speculum Historiale*, iv.128; and Anderson, *Alexander's Gate*, pp. 71–72.

44. Anderson, *Alexander's Gate*, p. 7; Westrem, "Against Gog and Magog," p. 55.

45. W.W. Skeat, ed., *Alexander and Dindimus*, EETS o.s.31 (1878; repr. London: H. Milford for Oxford University Press, 1930), p. xviii.

46. Wells (*A Manual of Writings in Middle English, 1050–1400* [New Haven: Connecticut Academy of Arts and Sciences, 1916], I.104) claims that "The letters are really a presentation of all the author can say on the old theme of the contrast and relative worth of the Active Life and the Contemplative Life," concluding that "The whole is cleverly manipulated so as to exhibit the excellent qualities of each of the opposing elements, without seeking to persuade to conclusion in favor of either." R.M. Lumiansky, in the revised manual, sees in the device of the letters "a balancing of aspects rather than an outright debate" (*A Manual of the Writings in Middle English, 1050–1500*, J. Burke Severs, gen. ed., Vol. I fasc. 1.5 [New Haven: Connecticut Academy of Arts and Sciences, 1967], 108). For Bennett, who goes on to observe that the poem "scarcely merits such ornament" as it receives in MS Bodley 264, see *Middle English Literature*, ed. and compl. Douglas Gray (Oxford: Clarendon, 1986), pp. 92–93. Turville-Petre's judgment that "It is a sad irony

that, of all the magnificent poems of the Revival, it was the incompetent and tedious *Alexander and Dindimus* alone on which scribe and illuminator lavished their professional attention" can be found in his *The Alliterative Revival* (Cambridge: D.S. Brewer, 1977), p. 43.

47. The manscript context of the poem is also of puzzling interest; the unique copy of *Alexander and Dindimus* was apparently added to the luxurious MS Bodley 264 ca. 1400 to supplement a mid-fourteenth-century copy of the French *Roman d'Alexandre*; it is followed by a French prose version of Marco Polo's *Li Livres du Grant Caam*, also added to the MS ca. 1400–10. For speculation about how and why the these additions were made, see my "Contextualizing *Alexander and Dindimus*," *YLS* 18 (2004): 81–106.

48. See Hanna's chapter on "Alliterative Poetry," in Wallace, ed., *The Cambridge History of Medieval English Literature*, p. 511. He continues, "Yet at the same time [the poems] worry and lament the burden of that consciousness, for they are oppressively aware of the futility of efforts at pursuing justice. Alliterative lords may conquer gloriously, but they never vanquish their own failure to operate without exploitation. For them, history is a longing for a new beginning, but a beginning which can never be disentangled from the preceding end, the tyranny inherent in rule."

49. "Langland's Lives: Reflections on Late-Medieval Religious and Literary Vocabulary," in *The Idea of Medieval Literature: New Essays on Chaucer and Medieval Culture in Honor of Donald R. Howard*, ed. James M. Dean and Christian Zacher (Newark: University of Delaware Press, 1992), p. 238.

50. *Piers Plowman* C.10.122–26, from Derek Pearsall, ed., *Piers Plowman: An Edition of the C-Text* (Berkeley: University of California Press, 1978).

51. In other versions of this correspondence, they do meet, and the confrontation is typically humiliating for Alexander; see George Cary, "A Note on the Mediaeval History of the *Collatio Alexandri cum Dindimo*," *Classica et Mediaevalia* XV (1954): 124–29, and n.67 below.

52. The *Collatio Alexandri cum Dindimo per Litteras Facta* is probably a product of the fourth century, although it was revised in the tenth for inclusion in the *Historia de Preliis*. The origin of the *Collatio* and its relation to later works about Alexander are discussed by Cary, *The Medieval Alexander*, p. 14, and Magoun, *Gests*, pp. 46–47; the *Collatio's* own sources in Greek writing are traced by J.D.M. Derrett, "The History of Palladius on the Races of India and the Brahmans," *Classica et Mediaevalia* 21 (1960): 64–99.

53. Magoun, *Gests*, p. 177; his edition usefully supplies the source passage from the *Historia de preliis* at the foot of each page.

54. *Wars* increases the deadly menagerie even further:

> Þan list þe lord on his lyfe haue with þa ledis spoken,
> Miȝt he haue won ouir þe water for wounding of bestis,
> As see-bule[s] & serpentis & soukand l[e]ches,
> Bathe eddirs & ascres & attirand wormes;
> Þire cocatricesse in creuessis þaire kindiles þai brede,

Scorpions many score, scautand neddirs,
And allway bot in A[ugust], as þe buk sais,
And saue þe iolite of Iuly þai iowke in þa strandis.

(*Wars*, 4324–31)

55. The English poet adds another interpreter to the poem's opening scene with the Gymnosophists, an episode I discuss below; Alexander questions them "By ludus of þe langage" (56).

56. All of Alexander's letters are described as bearing his seal, some of them twice; cf. ll. 182, 256, 817, 968, and 1085. Dindimus does not seem to have a seal, as doubtless befits his virtuous poverty, and the *Wars*-poet does not describe Alexander's letters as sealed. Presumably some sort of signet is what is meant; the letters depicted in some of the illustrations accompanying the poem do not show the hanging seals typical of letters patent. Four of the nine pictures associated with the poem show Alexander or Dindimus receiving letters from the hand of a messenger, and a fifth depicts Dindimus writing a letter while a courier waits to deliver it to Alexander. Moreover, Alexander is depicted as crowned but not armored, in contrast to the miniatures in the French *Roman*, where he often appears in armor and in battle.

57. Emily Steiner, *Documentary Culture and the Making of Medieval English Literature* (Cambridge: Cambridge University Press, 2003), p. 11.

58. This territory is thoroughly surveyed by Thomas Hahn in "The Indian Tradition in Western Medieval Intellectual History," *Viator* 9 (1978).

59. *Paradiso* 19.70, trans. Singleton. At the same time we have the example of Philippe de Mézières, whose crusading spirit was evidently untroubled by such advocacy for the Brahmans' salvation; in his *Epistre au roi Richart* of 1395 he decorates the "vergier delitable" with paintings of "les temps dores de ce monde, c'est assavoir la gracieuse policie des Bargamains et de leur roy." They represent a Golden Age whose peacefulness appeals to de Mézières, insofar as peace between England and France was a necessary precondition to a renewed crusade. See his *Letter to King Richard II*, ed. G.W. Coopland (Liverpool: Liverpool University Press, 1975), p. 128. De Mézières also encounters the Brahmans in his pilgrim guise in *Le Songe du Vieil Pelerin*, ed. Coopland (London: Cambridge University Press, 1969), I.224–27.

60. Skeat, because he compared the poem to a I³ recension of the *Historia de preliis*, attributed to the English writer many "poetical interpolations ad libitum" (p. xiii); Magoun, in contrast, calls it "an almost slavishly faithful translation" of the I² text (pp. 71–72).

61. Magoun, *Gests*, p. 195, supplies the Latin. The *Wars*-poet succumbs to a similar temptation at this point: "God seȝis oure saȝes for his son at in himselfe duellis; / For sekire god is þe son þat all oure sede loues, / And sothly, by þe same son, we seme him all like, / And all he sustayned of þat son þat any saule wildis" (4607–10). Cf. Duggan and Turville-Petre's note to these lines, p. 274. The anonymous translator of the *Polychronicon* in MS Harley 2261 also gives in here, rendering Higden's "Cum enim ipse verbum spiritus ac

mens sit" as "Then sithe þat word and son of God be a spiritte. . ." (*Polychronicon*, ed. Lumby, iii. 461). Trevisa resists the urge in his translation. Other examples of such Scriptural citation include a paraphrase of I Cor. 10:31 at ll. 358–62 (cf. Magoun's note, p. 236) and an allusion to one of the Beatitudes, Matt. 5:8, at ll. 625–26, where God "clepeþ to is joie / Clene-minded men þat meke ben founde."

62. Other examples of this habit are ll. 217–20, 325–30, 445, 784, and 1078–81.

63. In *Alexander and Dindimus*, the word "alowe" cannot have this precise theo-logical sense Whatley ("*Piers Plowman* B 12.277–94") attributes to Langland's usage, for the particular salvific grace to which Imaginatif alludes in *Piers* is beyond the imagination—or at least the vocabulary—of the poem's pagan protagonists. The word is certainly theologically inert in l. 259, where it translates "laudamus."

64. Cf. also ll. 508–15, at about the midpoint of Dindimus's letter: " ʒif þou our lif wole alowe and oure lawe use, / Hit schal þe profite, prince, whan þei pres faileþ./ Hit is noht long on us, lud, þei hit loþ seme, / For y have sent þe my sonde as þou þeiself bade. / But be þou nouht, bolde king, balful no tened / Þat þou miht trystli trye þe treweste lawe. / For we schulle minnge þe, man, swiche maner lorus / Þat þou miht lihtliche, lud, þe beste lawe kenne." In the Latin here, Dindimus claims that if Alexander chooses to follow the Brahman law, he will find it difficult, not profitable: "credimus tibi durum esse."

65. In fact, the original version of the correspondence with Dindimus was probably intended to be a defense of Alexander against Cynic or Stoic attacks on his character. See Cary, *Medieval Alexander*, pp. 13–14 and "Note," p. 125; Hahn, "Indian Tradition," pp. 219–20; and E. Lienard, "La Collatio Alexandri et Dindimi," *Revue Belge de Philologie et d'Histoire* 15 (1936): 819–38.

66. Seeing Dindimus as Alexander's foil rather than his philosphical instructor also helps explain the Brahmans' strongly negative rhetoric; their asceticism only makes sense as the opposite of the Greeks' alleged indulgence. Epistolary writing in general permits the kind of reciprocal self-definition in which each correspondent is defined in terms of the other, and this structure is elaborated into a style in *Alexander and Dindimus*. When Dindimus describes the habits of the Brahmans in terms of Western mores and culture, with references to practices familiar to both Alexander and the contempo-rary medieval reader, he emphasizes denial, absence, and negation:

> Hit is *no* leve in oure lawe þat we land erie
> Wiþ *no* scharpede schar to schape þe forwes,
> *Ne* sette solow on þe feld, *ne* sowe *non* erþe,
> In ony place of þe plow to plokke wiþ oxen,
> *Ne* in *no* side of þe se to saile wiþ nettus,
> Of þe finnede fihcs our fode to lacche.
> For to hauke *ne* hunte have we *no* leve,
> *Ne* foure-fotede best ferk to kille;
> *Ne* to faren in þe feld and fonde wiþ slyhþe

For to refe þe brod of briddus of hevene.
And whan we faren to fed we finde *no* faute:
We han so michel at þe mel þat we *no* more wilne.
Oþir goodis to gete give we *no* tente,
Ne oþir daynteys dere desire we *none*. . .

(293–306, emphasis added)

Wars and *Prose* also exhibit this stylistic trait, which greatly exaggerates the Latin's "apud nos illicitum est. . ." Even when describing the Brahmans' sufficiency of food in ll.303–04, Dindimus uses negatives: "we finde no faute"; "we no more wilne."

For observation on "epistolary discourse," I am indebted to Janet Altman, *Epistolarity:Approaches to a Form* (Columbus, OH: The Ohio State University Press, 1982), esp. chap. 4.

67. See Higden, *Polychronicon*, ed. Lumby, iii.472–79, and *Eulogium Historiarum*, i. 432–34. Neither of these moralizing historians goes quite as far as Johann Hartlieb, physician and diplomat for Duke Albrecht III of Bavaria. In his mid-fifteenth-century version of the story Alexander, despite being convinced of the truth of everything Dindimus says, nevertheless explains that he cannot emulate the virtuous Indian; he spends the rest of his life sadly regretting this decision, lamenting "O we, ach und we, dass ich der guten lere Dindimi nit gevolget hab!" See H. Becker, "Zur Alexandersage," *Zeitschrift für Deutsche Philologie* 23 (1891): 424–25.

68. *Winner and Waster*, ll.375, 378, from Stephanie Trigg, ed., *Wynnere and Wastoure*, EETS o.s. 297 (Oxford: Oxford University Press, 1990).

69. Hanna, "Alliterative Poetry," p. 505.

70. I borrow the phrase "thought experiment" from Elizabeth Fowler, who uses it to describe the *Knight's Tale* as an inquiry into the competing claims of dominion by conquest (represented by Theseus's triumph) and dominion by consent (represented by the exchange of marriage vows). See her "Chaucer's Hard Cases," in *Medieval Crime and Social Control*, ed. Barbara Hanawalt and David Wallace (Minneapolis: University of Minnesota Press, 1999), pp. 124–42. The idea that romance might interrogate lordship rather than simply celebrate its exercise is of course a virtual commonplace; cf. Chris Chism, *Alliterative Romance* (Philadelphia: University of Pennsylvania Press, 2002), p. 2: "Like their insular romance predecessors described by Rosalind Field and Susan Crane; [sic] these poems (1) investigate the historical antecedents of medieval structures of authority; (2) dramatize the questioning of cultural centers from outsider (or provincial) perspectives; and (3) centralize the historical exigencies of a world in flux rather than aiming primarily at more transcendent concerns with the afterlife. However, alliterative romances extend these themes, accentuating the mutually structuring oppositions between past and present, center and periphery, secular and religious trajectories, in order to integrate their reciprocal debts and interdependencies."

71. On the monastic roots of this division, see T.P. Dunning, "Action and Contemplation in *Piers Plowman*," in *Piers Plowman: Critical Approaches*,

ed. Hussey, pp. 213–25. There is, contemporary with *Alexander and Dindimus*, a wave of Latin treatises on the origins and preeminence of the life of regulars throughout England, advancing from Bury to Durham, St. Albans, Glastonbury and elsewhere. These texts, probably first written in response to the challenge of the friars (and later the Lollards), typically rely on scriptural precedent to buttress their claims on behalf of the monastic life. But *Alexander and Dindimus* cites Scripture only allusively and incidentally; it may be related to this literature of monastic self-scrutiny if, like other alliterative translations, it was originally a monastic production, but the relationship is a distant one. See Knowles, *Religious Orders in England II*, pp. 270–72; W.A. Pantin, "Some Medieval English Treatises on the Origins of Monasticism," in *Medieval Studies Presented to Rose Graham*, ed. Veronica Raffer and A.J. Taylor (Oxford: Oxford University Press, 1950), pp. 189–215; Pantin, "Two Treatises of Uthred de Boldon on the Monastic Life," in *Studies in Medieval History Presented to F.M. Powicke*, ed. R.W. Hunt, W.A. Pantin, and R.W. Southern (Oxford: Clarendon, 1948), pp. 363–85; Wendy Scase, *Piers Plowman and the New Anticlericalism*, (Cambridge: Cambridge University Press, 1989), pp. 88–94; and for Thomas Walsingham's contribution to this topic, Clark, "Thomas Walsingham Reconsdered," p. 851.

72. The other occurs in the challenge of the Gymnosophists cited above: "How miȝt þu kepe þe of sckaþe with skile and with trouþe / Aȝeins ryht to bireve rengnus of kingus?" (81–82). Alexander replies with his claim of divine commission. Dindimus once uses the phrase "to say þe truþe" (275), in the sense of "give an accurate account of"; it has no ethical weight.

73. The alliterative collocation "to sowe and to sette" mirrors Langland's "to sette or to sowe," a phrase that appears several times in his poem, and according to the *MED* nowhere else in the alliterative corpus. The phrase appears in Piers's first speech (A.6.29, B.5.541, C.7.186); in the opening of C's Pardon scene, where those who help Piers to plow are the first to be granted pardon (C.9.6; at B.7.6 Kane-Donaldson print "to erye or to sowe," but most MSS actually read "to erye or to sette or to sowe" or something very similar); and in two other places: A.7.127 and C.15.211. The *MED* (s.v. "setten," 3a & c) cites *Piers, Alexander and Dindimus, Dives et Pauper*, and a contemporary (1416) gardening text as its earliest examples.

Chapter 4   The Rhetoric of the Righteous Heathen

1. *Chronicles of the Revolution, 1397–1400: The Reign of Richard II*, trans. and ed. Chris Given-Wilson (Manchester: Manchester University Press, 1993), p. 200.

2. "Dominus de Cobham inprimis, praemisso longo sermone de malitia transacti temporis, dixit inter caetera, quod sub talibus Rege, ducibus, et rectoribus, conditio Anglorum facta fuit pejor quorumlibet conditionis ethnicorum, qui licet infideles sint ad fidem Christianam, et male creduli inter se, tamen vera loquuntur, vera faciunt, veraque fatentur. Sed Anglici, cum sint Christiani, et professores vertitatis esse debant, et in ipsa

perseverare, metu amissionis suae substantiae temporalis, metu detrusionis in exilium, metu denique mortis, qui in constantes etiam posset cadere, nullibi in agendis verum facere vel verum dicere sub talibus gubernatoribus fuerunt ausi. . . ." From Walsingham's *Annales Ricardi Secundi*, in *Johannis de Trokelowe Anon. Chronica et Annales*, ed. H.T. Riley, Rolls Series 28:3 (London: Longmans, Green, Reader and Dyer, 1866), p. 306. Trans. Given-Wilson, *Chronicles of the Revolution*, p. 204.

John de Cobham inherited his title on his father's death in 1355 and must have been quite old in 1399; Walsingham says that the king's decision to banish Cobham rather than execute him was due to his age: "Rex tamen, concessa seni, quam non optavit, venia, sive vita, misit eum ad insulam de Gerneseya in exilium" (*Ypodigma Neustriae*, ed. H.T. Riley, Rolls Series 28: 7 [London: Longman, 1876], p. 379). Cobham had acquired considerable diplomatic and adminstrative experience under Edward III and Richard II, but his service on the commission to which the Lords Appellant brought their charges in 1386–88 made him a target of Richard's vengeance in 1397, despite the fact that, according to Gower's *Cronica tripertita*, he had by this time retired to a Carthusian monastery (*Cronica tripertita* II: 212–32, in Gower, *Complete Works*, ed. G.C. Macaulay, 4 vols. [Oxford: Clarendon Press, 1899–1902], iv. 326). After Henry's accession, Richard's persecution of Cobham became one more item on the laundry list of the *Record and Process* composed to justify the Lancastrian usurpation. He died in 1407–08 (possibly close to 100, the *DNB* speculates); his granddaughter Joan was his heir, and it was by virtue of marriage that her fourth husband (of five), Sir John Oldcastle, was known as Lord Cobham.

Cobham was doubtless acquainted with both Chaucer and Gower; he and Chaucer served on the same commission of the peace in Kent between 1385 and 1389, while he and Gower were involved in a the estate-purchase dispute known as the "Septvuans affair" in the previous decade. For Cobham's life, see *DNB* IV 611–12 and J.G. Waller, "The Lords of Cobham, their Monuments, and the Church," *Archaeologica Cantiana* XI (1877): 49–112, esp. 70–99. On Chaucer's service as justice of the peace for Kent, see Derek Pearsall, *The Life of Geoffrey Chaucer* (Oxford: Blackwell, 1992), pp. 205–06, and Martin M. Crow and Clair C. Olson, *Chaucer Life-Records* (Oxford: Clarendon Press, 1966), pp. 348ff. Cobham was also involved in the Scrope–Grosvenor controversy in which Chaucer gave testimony in 1386, and was traveling to France for peace and marriage negotiations at the same time Chaucer was doing the same (1377–1381), though no extant documents put them on the same commission; see *Life-Records*, pp. 360n and 53n. For Gower and Cobham, see John Fisher, *John Gower: Moral Philosopher and Friend of Chaucer* (New York: New York University Press, 1964), pp. 51–54.

3. Still another fourteenth-century example, this time in a pastoral setting, can be found in *The Book of Vices and Virtues* (ed.W. Nelson Francis, EETS o.s. 217 [Oxford: Oxford University Press, 1942; repr. 1968], p. 124): "A, God, how we schulde be a-ferd, whan þei þat weren heþen and wiþ-out any lawe y-write, þat wisten no þing of þe verray grace of God ne of þe Holi Gost, and ȝit

clomb þei vp to þe hill of parfiȝtnesse of lif bi strengþe of here owne vertue, and deyned not to loke on þe world; & we þat ben cristene and hadde þe grace and þe bileue veraliche and conen þe comaundements of God and han þe grace of þe holi Gost, ȝif we wolen, and more we myȝt do profiȝt in on day þan þilke myȝten in an hole ȝere, we lyuen as swyn here byneþe in þis grottes of þis world! And þerfore seiþ seynt Poule þat þe heþen þat ben wiþ-oute lawe, at domes day schul jugge vs, þat han þe lawe & don it nouȝt. . ." A similar passage appears in *Dan Michel's Ayenbite of Inwit* (ed. Pamela Gradon, EETS o.s. 278 [Oxford: Oxford University Press, 1979], p.126), since both are translations of the French Dominican Lorens d'Orléans's *Somme le Roi* of ca. 1279; the *Ayenbite* dates from 1340, the *Book* perhaps a bit later (Francis, p. lxviii). The scriptural reference is to Romans 2: 25–27: "those who are physically uncircumcised but carry out the law will pass judgement on you, with your written law and circumcision, who break the law." Minnis cites the appearance of the cliché in Augustine, Gregory the Great, and (citing Gregory) John of Wales; see his "From Medieval to Renaissance? Chaucer's Position on Past Gentility," *Proceedings of the British Academy* LXXII (1986): 213–16, as well as *Chaucer and Pagan Antiquity* (Cambridge: D.S. Brewer, 1982), pp. 1–6, 31, 62–63, and "Looking for a Sign" in *Essays in Ricardian Literature in Honour of J.A. Burrow*, ed. Minnis, Charlotte Morse, and Thorlac Turville-Petre (Oxford: Clarendon, 1997), pp. 147–48.

4.   "And [Cobham] went on to say that it was clear to him that since the king himself, who was the chief and foremost among so many worthless and evil counsellors, had quite rightly been deposed and punished for this sort of crime, so also should those who had encouraged or incited or persuaded him to commit such deeds be arrested and made to suffer whatever punishment it was thought that their wicked counsel deserved, which should be decided by sober deliberation in a council of the realm" (". . .et in progressu dixit, visum fore, ut cum Rex ipse, qui fuerat caput et capitaneus tot nequam malorum consiliariorum, sit depositus et punitus merito propter hujusmodi scelera, quod illi qui haec suaserunt sibi, instigarunt, et quodammodo ad talia perpetranda coegerunt eum, sub detentione servarentur, et, secundum quod sanum consilium regni dictaverat, meritas sui nequam consilii poenas luant"). Trans. Given-Wilson, p. 204; *Trokelowe*, pp. 306–07. One might also argue that Cobham's use of the righteous heathen contrast helps to obscure the otherwise obvious self-regard in his observation: here is one Christian who has suffered exile, loss of goods, and fear of death but who is not reluctant to speak the truth.

5.   Minnis, "Looking for a Sign," p. 156.

6.   Why Arcita's scene is transposed to the end of *Troilus* and not used in the *Knight's Tale* is easy enough to speculate about; in *Troilus* the episode helps to construct the Christian frame to the poem, whereas in the *Knight's Tale*— narrated as part of the Canterbury pilgrimage—it would tend to challenge the frame or even break it. Moreover, Arcite's celestial installation there would confirm Theseus's claims that "a man hath moost honour / To dyen in his excellence and flour" (I.3046–47), and thus undermine the way

Chaucer casts doubt on Theseus's Neoplatonic "Firste Moevere" speech at the end of the tale. What can't be known for sure, on the other hand, is whether the poem referred to in the *Legend of Good Women* as "the love of Palamon and Arcite / Of Thebes" (*LGW* F 408–09) originally included a version of the apotheosis.

Henry Ansgar Kelly discusses the mechanics of the insertion of the *Teseida* passage in his *Chaucerian Tragedy* (Woodbridge, Suffolk: D. S. Brewer, 1997), pp. 132–34.

7. In fact Arcite does not so much enumerate his virtues as list the many sins he has failed to commit, clearing himself of the guilt associated with previous generations of Thebans—Cadmus, Oedipus, Creon, etc. See *The Book of Theseus: Teseida delle Nozze d'Emilia*, trans. Bernadette Marie McCoy (New York: Medieval Text Association, 1974), X.94–99.

8. Donaldson's account of the ending of *Troilus and Criseyde* describes the anti-climactic structure of these stanzas, which begin in "emphatic disgust" and end in "exhausted calm" (*Speaking of Chaucer* [New York: W.W. Norton, 1970], pp. 97–99). Their rhetorical effect is paradoxical as well, since their attempt to reject Troilus's paganism and condemn his "fyn" nevertheless directs our attention back to that topic, and its optimistic obscurity. Troilus's own sudden and unexplained Neoplatonic insight—he "fully gan despise / this wrecched world, and held al vanyte / To respect of the pleyn felicite / That is in hevene above" (5.1816–19)—utterly preempts our own, or that which is wished for on our behalf in the ensuing stanzas. However strongly one sees *Troilus and Criseyde* as bearing a Boethian moral, Troilus, unlikely enough, gets there first. Some critics find that he has taken an unacceptable shortcut, e.g. Alan Gaylord, who argues that Troilus's final vision "is more like a quick peek in the back of the book than a completed lesson" ("The Lesson of the *Troilus*: Chastisement and Correction," in *Essays on Troilus and Criseyde*, ed. Mary Salu [Cambridge: D.S. Brewer, 1979], p. 25).

9. Donaldson takes as given Troilus's status as a virtuous pagan: "The three stanzas describing Troilus's afterlife afford him that reward which medieval Christianity allowed to the righteous heathen" (*Speaking of Chaucer*, p. 96). See also Barry Windeatt, *Oxford Guides to Chaucer: Troilus and Criseyde* (Oxford: Oxford University Press, 1992), p. 211: "Against a background of some controversy in Chaucer's day over the possibility of salvation for virtuous pagans, it is especially apposite that in the Neoplatonic and Hermetic tradition of the *ogdoad* souls possessed of gnosis return to the eighth sphere at death, and it may well be that by a reference to the eighth sphere Chaucer intended to invoke the possibility allowed by this tradition of a suitable point to which the soul of Troilus might ascend, although he is careful not to say that this was the permanent resting-place of the hero's soul." Kelly, *Chaucerian Tragedy*, pp. 132–36, takes up the question as well, remarking (p. 132) that "We saw that Pandarus once told Troilus, in an 'applied' context, 'Thow shalt be saved by thi feyth in trouthe' (2.1503), and on another occasion he pointed out to him that if he died a martyr he would go to heaven (4.623)."

10. Wetherbee, *Chaucer and the Poets: An Essay on Troilus and Criseyde* (Ithaca: Cornell University Press, 1984), p. 241. See also Minnis, *Chaucer and Pagan Antiquity*, pp. 105–07, and T.P. Dunning, "God and Man in *Troilus and Criseyde*," in *English and Medieval Studies Presented to J.R.R. Tolkien on the Occasion of his Seventieth Birthday*, ed. Norman Davis and C.L.Wrenn (London: George Allen & Unwin Ltd., 1962), pp. 164–68. Wetherbee compares Troilus to the *Thebaid*'s Menoeceus and finds him "poised for flight" at the end, a flight that is thus explained if not anticipated. At the same time, though, Wetherbee includes as another analogous figure the Statius imagined by Dante in *Purgatorio* XXI, who describes how purified souls are "surprised" by the freedom of the will that accompanies the end of their purgation (pp. 228–35). Chaucer may have found in this episode an implicit precedent for his own narrative surprise, given that Dante's representation of Statius as a pagan who had covertly converted to Christianity was apparently his own creation (and thus comes as a surprise to his readers); moreover Statius himself is surprised to discover that he is in the presence of the Vergil he has just been praising— so surprised, in fact, that he tries to embrace Vergil, forgetting that they are both mere shades. The tableau—a putatively Christian poet trying but failing to embrace a pagan poet he admires—is an undeniably poignant one.

11. As Steven Justice has observed, the Nun's Priest's brief discussion of "symple necessitee" and "necessitee condicioneel" recapitulates Troilus's meditations in book 4 (*Writing and Rebellion* [Berkeley: University of California Press, 1994], pp. 219–21). In fact the narrator's interruption (*NPT* 3226–66) is full of recollections of *Troilus and Criseyde*, not only Boece (3242) and the fall of Troy (3228–29) but also the recourse to *auctors* (3263) and the defusing of a potentially antifeminist moral (3260–66) similar to that attempted at the end of the earlier poem ("Beth war of men, and herkneth what I seye!" [5.1785]).

    On Chaucer's likely knowledge of *Piers Plowman*, see e.g. J.A.W. Bennett, "Chaucer's Contemporary," in *Piers Plowman: Critical Approaches*, ed. Hussey, pp. 310–24; George Kane, *Chaucer and Langland: Historical and Textual Approaches* (Berkeley: University of California Press, 1989); and my own "Chaucer Reading Langland: *The House of Fame*," *SAC* 18 (1996): 3–23. For Chaucer's use of Holcot, see Robert A. Pratt, "Some Latin Sources of the Nonnes Preest on Dreams," *Speculum* 52 (1977): 538–70, and for a summary account of these mid-century theological controversies in England see Minnis, *Chaucer and Pagan Antiquity*, pp. 55–60, and Leff, *Bradwardine and the Pelagians* (Cambridge: Cambridge University Press, 1957).

12. Windeatt, *Oxford Guides*, p. 231. On the topic of conversion in the poem, and the contrast between Troilus's sudden Pauline *volte-face* and Criseyde's more elaborate Augustinian process, see Dabney Anderson Bankert, "Secularizing the Word: Conversion Models in Chaucer's Troilus and Criseyde," *ChR* 37 (2003): 196–218.

13. See Minnis, *Chaucer and Pagan Antiquity*, esp. pp. 93–100. But note that he too recognizes the virtuous pagan analogy; describing the hymns of book 3, he observes "By rationalizing from his own experience and 'doing what was in him', Troilus has transcended the polytheism and fatalism which he

espouses elsewhere in the poem" (100). For a more critical assessment of Troilus's philosophical achievements, see John Fleming, *Classical Imitation and Interpretation in Chaucer's Troilus* (Lincoln: University of Nebraska Press, 1990).

14. R.K. Root, in his notes to the poem's epilogue, also points to these lines in the *Parliament*, as well as to the *Somnium Scipionis*; see his *The Book of Troilus and Criseyde* (Princeton: Princeton University Press, 1926; repr. 1945), p. 562. Monica McAlpine (*The Genre of Troilus and Criseyde* [Ithaca: Cornell University Press, 1978], pp. 179–80) cites the *Somnium*, but argues that the allusion "limits and qualifies the authority of Troilus's vision" because it ties the episode to a "seriously incomplete" pagan wisdom. On the differences between the narrator of the dream-visions and the narrator of *Troilus and Criseyde*, see David Lawton, *Chaucer's Narrators* (Woodbridge, Suffolk: D.S. Brewer, 1985), pp. 77–90.

15. Cf. Wetherbee, *Chaucer and the Poets*, p. 234: "Despite the richness of Chaucer's account of Troilus's enlightenment, with its echoes of the *Sompnium Scipionis* and the *Paradiso*, we must recognize that he is actually suspended in a spiritual void, that there is no category of religious experience to which we can confidently refer his spiritual journey." If we think of the righteous heathen context as a literary category as much as a religious one, our confidence in Chaucer's benign intentions can increase considerably.

For a less sanguine reading of this passage in the *Parliament*, focusing on its hegemonic potential, see Watson, "Visions of Inclusion: Universal Salvation and Vernacular Theology in Pre-Reformation England," *JMEMS* 27 (1997): 172–73.

16. Donaldson, *Speaking of Chaucer*, p. 91.

17. Lee Patterson, *Chaucer and the Subject of History* (Madison, WI: University of Wisconsin Press, 1991), p. 154.

18. For a history (and reconsideration) of this critical assumption, see A.C. Spearing, "A Ricardian 'I': The Narrator of 'Troilus and Criseyde'," in *Essays in Ricardian Literature*, ed. A. Minnis et al., pp. 1–22.

19. The treatment accorded to the ladies is of course considerably more problematic than this, given Criseyde's virtual absence from the epilogue and the end of the poem generally. She last appears in her own words at 5.1631, prophetically wishing for Troilus that "God have yow in his grace!"; a hundred lines later she is the object of Pandarus's abiding hate, and a hundred lines after that the object of a preposition: "And thus bigan his lovyng of Criseyde, / As I have told, and in this wise he deyde" (5.1833–34). Progressive erasure, rather than apotheosis, characterizes Criseyde's final appearances. Slightly more charitably one could say that her apotheosis consists of her replacement by the "mayde and moder" Mary in the last line of the poem; cp. E.T. Donaldson: "The poem has concerned a mortal woman whose power to love has failed, and it ends with the one mortal woman whose power to love is everlasting" (*Speaking of Chaucer*, pp.100–01). Much more than in the *de casibus* tradition—which thanks to Boccaccio eventually develops its *de claris mulieribus* counterpart—the discourse of the virtuous pagan is a masculine one. On Criseyde's fate see Carolyn Dinshaw, *Chaucer's*

Sexual Poetics (Madison: University of Wisconsin Press, 1989), pp. 52–64, and Gayle Margherita, *Romance of Origins: Language and Sexual Difference in Middle English Literature* (Philadelphia: University of Pennsylvania Press, 1994), pp. 100–12.

20. Geraldine Heng, *Empire of Magic: Medieval Romance and the Politics of Cultural Fantasy* (New York: Columbia University Press, 2003), p.116.

21. All citations of the poem are drawn from Mary Hamel, ed., *Morte Arthure: A Critical Edition* (New York: Garland Publishing, 1984), here ll. 1, 4–6.

22. On the sources of the *Morte*, see Hamel, pp. 34–53; for the *Ferumbras* connection, see John Finlayson, "The Alliterative *Morte Arthure* and *Sir Ferumbras*," *Anglia* 92 (1974): 380–86. The text of *Ferumbras* was edited by Sidney J. Herrtage, EETS o.s. 34 (London: Kegan Paul, Trench Trübner & Co., 1879). As Lee Patterson notes (*Negotiating the Past*, pp. 220–21) the Alexander tradition is constantly alluded to in the *Morte*, not only here but in the earlier vowing scene and in the later dream of Fortune; as we have seen, it is a tradition that has its own provocative connections to the discourse of the righteous heathen. Moreover, the embedded nature of the scene recalls Sir John's conversation with the Sultan, an incident borrowed from one source (Caesarius of Heisterbach) and inserted into a translation of another (the *De statu Saracenorum*). See chapter two.

23. Chism, *Alliterative Revivals* (Philadelphia: University of Pennsylvania Press, 2002), p. 221; see also e.g. Heng, *Empire of Magic*, p. 156, and *Morte Arthure*, ed. John Finlayson (Evanston, IL: Northwestern University Press, 1967), pp. 10–11.

24. Patterson, *Negotiating the Past*, p. 220. As many critics have observed, Priamus's heritage obviously connects him with Arthur, who will soon take his place among the Worthies in his dream of Fortune. Priamus's particular forbears are the non-Christian Worthies, two of the three pagans (Julius Caesar being omitted, presumably, so as to minimize Priamus's connection with Rome) and two of the three Hebrews (David's absence perhaps being explained by his genealogical connection to Christ, a line of descent into which it would be difficult to fit Priamus). His invented name, of course, also associates him with Troy, and recalls Langland's adoption of the Latin variant of Trajan's name, "Troianus." On that form of the name, see the definitive study by Siegfried Wenzel, "Langland's *Troianus*," *YLS* 10 (1996): 181–85.

   Hamel argues that Priamus is not a pagan but rather a Greek Christian who would have been considered schismatic by the Latin Church, and that his reference to "thy Cryste" thus implies difference in doctrine rather than faith; see her edition of the poem, p. 40, and "The 'Christening' of Sir Priamus in the Alliterative *Morte Arthure*," *Viator* 13 (1982): 295–307. I find this too literal a reading of the details of Priamus's heritage. Heng also considers Priamus Christian, though not necessarily a Latin Christian, suggesting that Priamus's two invocations of St. Peter "would be substantially meaningless for a pagan, Muslim, or Jew" (*Empire of Magic*, pp. 155–56). But the first of these, "Petire!" (l. 2646), is clearly a commonplace interjection without theological force—Chaucer's eagle uses it this way in the *House of*

Fame (l. 2000)—and the second reference ("so helpe Seynt Peter!" l.2742) is actually spoken by Gawain, not Priamus.

25. Patterson, *Negotiating the Past*, p. 220.

26. Patterson, *Negotiating the Past*, p. 221, 226. The sacramental gesture in the scene is actually made by Priamus, who offers to heal Gawain's wound with the contents of a golden vial "þat is full of þe flour of þe four well / Þat flowes oute of Paradice" (2705–06). Malory's later version of the Priamus scene, adapted from the *Morte*, offers an instructive contrast, and suggests not only that he specifically recognized the earlier poem's omission of the conventional conversion but that the situation caused him a certain amount of anxiety. Not only does his Priamus specifically ask for baptism, but Malory also depicts the sacrament itself a few pages later, inventing a scene in which Gawain brings Priamus to Arthur, who "in haste crystynde hym fayre and lette conferme hym Priamus, as he was afore," making him a knight of the Round Table. See Sir Thomas Malory, *Works*, ed. Eugène Vinaver, 2nd edn. (Oxford: Oxford University Press, 1977), pp. 137 and 143. The Alliterative *Morte* and the Winchester text of Malory seem to be the chivalric equivalents of *Piers Plowman* and *St. Erkenwald* when it comes to the necessity of baptism.

27. On the date of the poem see Hamel's edition, pp. 53–58.

28. Work that takes note of the structural parallels in these scenes—typically the first and the last—includes Heng, *Empire of Magic*; Hamel, *Morte Arthure*; Patterson, *Negotiating the Past* (esp. p. 225 n.67); George Keiser, "The Theme of Justice in the Alliterative *Morte Arthure*," *Annuale Medievale* 16 (1975): 94–109; Jean Ritze-Rutherford, "Formulaic Macrostructure: The Theme of Battle," in *The Alliterative Morte Arthure: A Reassessment of the Poem*, ed. Karl Heinz Göller (Cambridge: D.S. Brewer, 1981), pp. 83–95; Jan Ziolkowski, "A Narrative Structure in the Alliterative Morte Arthure, 1–1221 and 3150–4346," *ChR* 22 (1988): 234–45; Lesley Johnson, "King Arthur at the Crossroads to Rome," in *Noble and Joyous Histories: English Romances, 1375–1650*, ed. Eiléan Ní Cuilleanáin and J.D. Pheifer (Dublin: Irish Academic Press, 1993), pp. 87–111; Robert Warm, "Arthur and the Giant of Mont St. Michel: The Politics of Empire Building in the Later Middle Ages," *Nottingham Medieval Studies* 41 (1997): 57–71; and Jeffrey Jerome Cohen, *Of Giants: Sex, Monsters, and the Middle Ages* (Minneapolis: University of Minnesota Press, 1999), pp. 152–58.

29. This category would include Patterson, Chism and Heng, and should probably be traced back to Larry Benson's exemplary New Critical analysis of the poem's late-medieval "gothic" aesthetic in "The Alliterative *Morte Arthure* and Medieval Tragedy," *Tennessee Studies in Literature* 11 (1966): 75–86. Benson writes, "The tension in a work like the *Morte Arthure* is thus not between good and evil, between the 'excess' of earthly kingship and the virtue of renunciation; the tension is between two goods, between the Christian detachment that is necessary for ultimate happiness even on the earth and the complete engagement with an earthly ideal that is necessary for heroism" (pp. 80–81).

30. The battle with the giant combines two such encounters in the sources, while the knight-pilgrim Cradoke's message is typically delivered by an

anonymous and thematically uninteresting messenger. See John Finlayson,
"Arthur and the Giant of St. Michael's Mount," *MAE* 33 (1964): 112–20, and
Hamel, *Morte Arthure*, p. 368 n.3509. The details of Arthur's meeting with Sir
Cradoke make it strikingly parallel to Gawain's with Priamus. Arthur is in
Priamus's position—alone, gaudily dressed, and standing still. He meets someone bound for Rome (as is Gawain, since he is part of Arthur's army), whom
he warns about the troops nearby (an element that's out of order, it should be
noted); the other—Cradoke—is like Gawain defiant that his way out will not
be blocked. There is a contrast in the messages delivered, of course; Cradoke's
is to fear the future represented by Mordred and the next generation, while
Priamus's message is that the past need not be feared, since it can be domesticated through an acknowledgment of chivalric continuity.

31. Macaulay, *Works* II. 515. Citations from the *Confessio amantis* will be taken
from this edition, where it appears in volumes two and three. For a contrary
point of view, see R.F. Yeager, *John Gower's Poetic: The Search for a New Arion*
(Cambridge: D.S. Brewer, 1990), pp. 170–87.

32. Larry Scanlon, *Narrative, Authority and Power: The Medieval Exemplum and the
Chaucerian Tradition* (Cambridge: Cambridge University Press, 1994),
pp. 264–65. Scanlon offers a subtle reading of the tale's politics on pp. 263–67,
drawing for a different purpose on many of the plot details I also use here.
On this tale see also Patrick J. Gallacher, *Love, the Word and Mercury: A Reading
of John Gower's Confessio amantis* (Albuquerque: University of New Mexico
Press, 1975), pp. 120–24; J.D. Burnley, *Chaucer's Language and the Philosophers'
Tradition* (Cambridge: D.S. Brewer, 1979), pp. 145–47; and Kurt Olsson,
John Gower and the Structures of Conversion (Cambridge: D.S. Brewer, 1992),
pp. 102–06.

33. The parallels continue past the point of Constantine's cure; whereas earlier
(2. 3217–19) he had sent out "with lettres and with seales" his order that the
children be collected, after his conversion he "sende anon his lettres oute /
And let do crien al aboute / Up peine of deth that noman weyve / That he
ne baptesme receive" (2.3467–70). For Gower's use of the Sylvester story in
the late English poem "In Praise of Peace," see my "The Lancastrian Gower
and the Limits of Exemplarity," *Speculum* 70 (1995), pp. 568–70.

For the resonance of the word, "enformacioun," and its various cognates
in the poem, see James Simpson, *Sciences and the Self in Medieval Poetry: Alan
of Lille's Anticlaudianus and John Gower's Confessio amantis* (Cambridge:
Cambridge University Press, 1995), pp. 1–15.

34. As Macaulay notes (II.492), in most versions of the story Constantine meets
the weeping mothers as he is riding to the Capitol for his bloodbath;
Gower's version omits this detail and places everyone in the palace, making
his Constantine a little less like Trajan than he might be. Whether this is an
intentional alteration or the result of Gower's adherence to a particular
source cannot be finally determined; on the possible sources of the tale see
Larry Scanlon, *Narrative, Power, Authority*, pp. 263–4 and 264n.27.

35. The phrase is not original with Minnis—Beryl Smalley has some claim to
it—but he gives a good summary account of its application to Gower: "The

notion of good and bad behaviour 'in general' permeates. . .the *Confessio amantis* of John Gower (never, to my knowledge, accused of being a humanist). Because Christians and pagans share certain moral standards (however much they may differ in other areas), Gower's framework of the Seven Deadly Sins is by no means inappropriate in a work that orchestrates exempla from sources both 'ancient' and 'modern,' both classical and medieval. Ovid and the other heathen authorities (as interpreted in the Middle Ages) reached out toward Christian schemes of virtues and vices; medieval Christians who described the moral principles and patterns of their religion looked back to the pagan past for anticipations and confirmations of their present-day beliefs." ("From Medieval to Renaissance?", p. 214.) For Smalley's fourteenth century "classicizing friars"—Holcot, Ridewall, Waleys, and others—see her *English Friars and Antiquity in the Early Fourteenth Century* (Oxford: Blackwell, 1960).

36. See Scanlon, *Narrative, Power, Authority*, pp. 257ff.

37. Simpson, *Sciences and the Self*, p. 297; Diane Watt, *Amoral Gower: Language, Sex and Politics* (Minneapolis: University of Minnesota Press, 2003), pp. 114–18. The inconsistency charge is made by G.H.V. Bunt in "Exemplum and Tale in John Gower's 'Confessio amantis,' " in *Exemplum et Similitudo: Alexander the Great and other heroes as points of reference in medieval literature*, ed. W.J. Aerts and M. Gosman, Mediaevalia Groningana fasc. 8, (Groningen: E. Forsten, 1988), pp.145–57.

38. He claims that Gower and Lydgate, "weary of war and of conquerors, represent the last degradation of the material and the attacks used by the philosophers of antiquity" (*Medieval Alexander*, 257), sentiments that may have as much to do with the postwar genesis of Cary's thesis as with sensitivity to Gower's pacifism. On Gower and war, see R.F. Yeager, "*Pax Poetica*: On the Pacifism of Chaucer and Gower," *SAC* 9 (1987): 97–121.

39. See Cary, *Medieval Alexander*, 35–36, and George L. Hamilton, "Studies in the Sources of Gower," *JEGP* 26 (1927): 491–500. F. Pfister, "Auf den Spuren Alexanders des Grossen in der älteren englishen Literatur," *Germanisch-Romanische Monatsshchrift* 16 (1928): 85–86, disputes Hamilton's claim that Gower used the J2 *Historia*, arguing instead for the J1 recension; he also suggests Valerius Maximus as the source of the Diogenes tale in Book 3, which was widely known.

40. The *Secretum* was accepted as both historical and Aristotelian in the Middle Ages and circulated very widely. For Gower's use of it, see M.A. Manzalaoui, " 'Noght in the Registre of Venus': Gower's English Mirror for Princes," in *Medieval Studies for J.A.W. Bennett (Aetatis Suae LXX)*, ed. P.L. Heyworth (Oxford: Clarendon Press, 1981), pp. 159–83; George L. Hamilton, "Some Sources of the Seventh Book of Gower's *Confessio amantis*," *Modern Philology* 9 (1912): 323–46; and A.H. Gilbert, "Notes on the Influence of the *Secretum Secretorum*," *Speculum* 3 (1928): 84–98. Although it was ultimately based on an eighth-century Arabic source, the *Secretum* was twice translated into Latin (in the mid-twelfth and early thirteenth centuries) and survives in over 500 manuscripts in that language alone. At least thirteen Middle English versions

from the fifteenth century are known, including a rhymed translation begun by John Lydgate (and finished by Benedict Burgh). For a brief description of the history of this text, see M.A. Manzalaoui, ed., *Secretum Secretorum: Nine English Versions*, EETS o.s. 276 (Oxford: Oxford University Press, 1977), pp. ix–xvi; the editor treats the Arabic text and its backgound in "The Pseudo-Aristotelian *Kitab Sirr al-Asrar*: Facts and Problems," *Oriens* 23–24 (1974, for 1970–71): 148–257. English versions can be found in Manzalaoui, *Secretum Secretorum*; Robert Steele, ed., *Three Prose Versions of the Secreta Secretorum*, EETS e.s. 74 (London: K. Paul, Trench, Trübner, 1908); and John Lydgate and Benedict Burgh, *Secrees of Old Philosoffres*, ed. Steele, EETS o.s. 66 (London: K. Paul, Trench, Trübner, 1894). Steele prints the Latin "vulgate" version in *Opera hactenus inedita Rogeri Baconi*, fasc. 5 (Oxford: Clarendon Press, 1920). Judith Ferster discusses the various paradoxes of the text and its reception in England and Ireland in *Fictions of Advice: The Literature and Politics of Counsel in Late Medieval England* (Philadelphia: University of Pennsylvania Press, 1996), pp. 39–66.

41. On this approach to Gower's two languages in the *Confessio* see e.g. R.F. Yeager, "English, Latin, and the Text as 'Other': The Page as Sign in the Work of John Gower," *Text* 3 (1987): 251–67.

42. The messenger knight is apparently Gower's invention; on the sources of this tale, see more recently Mayasoshi Ito, "Gower's 'Diogenes and Alexander' and Its Philosophic-Literary Tradition," *Poetica* (Tokyo) 16 (1983): 66–77. Watt, *Amoral Gower*, p. 118, also notes that "Alexander demonstrates his sagacity and self-restraint" in the Diogenes episode.

43. Winthrop Wetherbee, "John Gower," in *The Cambridge History of Medieval English Literature*, ed. David Wallace (Cambridge: Cambridge University Press, 1999), pp. 601–02.

44. Manzalaoui, " 'Noght in the Registre of Venus,' " p. 162.

45. On the education of Alexander by Aristotle, see Peter Green, *Alexander of Macedon, 356–323 B.C.: a historical biography* (Berkeley: University of California Press, c1991). Once again, it is possible to argue for a particularly keen English interest in this aspect of Alexander's career. The *Wars*-poet elaborates on the topic in these words:

> Þan was he lede furthe belyfe to lere at þe scole,
> As sone as to þat sapient himself was of elde,
> Onane vnto Arystotill, þat was his awen maistir,
> And one of þe coronest clerkes þat euir knew lettir.
> Þan was he broȝt to a benke, a boke in his hand,
> And faste by his enfourme[r] was fettild his place.
> For it come noȝt a kyng son, ȝe knowe wele, to sytt
> Doune in a margon & molle emange othire schrewis.
> Sone wex he wittir & wyse & wondir wele leres,
> Sped him in a schort space to spell and to rede,
> And seþen to gramere he gase, as the gyse wald,
> And þat has he all hale in a hand quyle.

In foure or fife ȝere he ferre was in lare
Þan othire at had bene þare [ell]euyn wyntir;
Þat he suld passe him in þat plite vnpussible semed,
Bot at god will at gaa furth, qua may agaynstande?

(*Wars* 621–36)

As the editors of the poem have made clear, the English poet has here expanded his single-sentence source into two dozen lines, embellishing the account of Alexander's academic prowess and adding not only the suggestion of divine influence on his grades (634–36) but also the name of his instructor—Aristotle's name is not mentioned in the Latin version for four more chapters. See *Wars*, p. 194 n.621. The editors quote the *Historia de Preliis*:"In scolis itaque ubi sedebat, pugnabat cum sociis tam in litteris quam in loquelis et velocitate optinens principatum." These battles with his schoolmates are described in *Wars*, pp. 637–44, and are said to take place only in the absence of schoolmaster Aristotle.

46. Other stories in book 7 elicit Alexander's presence in the background: Darius, who appears in the tale of "King, Wine, Woman and Truth" at VII. 1783–1984, was Alexander's best-known foe; Diogenes, his antagonist from book 3, appears in the "Tale of Diogenes and Aristippus" at VII. 2217–2320, playing essentially the same role as before and thus forcefully recalling his earlier encounter with Alexander. Alexander appears *in propria persona* only once in book 7, and then only in manuscripts of the so-called second recension of the *Confessio*; in a short exemplum drawn from Valerius Maximus (so says the marginal gloss), Alexander plays the role of judge, as he had with the Pirate in book 3.

Of Alisandre in his histoire
I rede how he a worthi knight
Of sodein wraththe and nought of right
Forjugged hath, and he appeleth.
And with that word the king quereleth,
And seith "Non is above me."
"That wot I wel my lord," quoth he,
"Fro thy lordschipe appele I noght,
But fro thy wraththe in al my thought
To thy pitee stant myn appeel.["]
The kyng, which understod him wel,
Of pure pitee yaf him grace.

(VII. 3168*–79*)

It is tempting to take this passage as emblematic of Gower's general treatment of Alexander; the anecdote recognizes the king's faults (here his propensity to fall into "sodein wraththe"), but his character proves ultimately recuperable, susceptible to an entirely reasonable appeal to his pity and willing, while most in control of his faculties, to exercise grace and benevolence. In the same way is the Alexander of the *Confessio* susceptible to a kind of recuperation; the exemplum-Alexander of book 3 is transcended by the exemplary Alexander of book 7. His faults disappear; his wars and death are forgotten and his reputation is transformed, starting at the end of book 6, into that of "the worthiest of alle kiththe." But the story of Alexander's pity is awkwardly self-referential; it breaks the fiction of the seventh book to have Alexander appear as an example for himself, and the tale is thus excluded from all but five of the manuscripts of the *Confessio*.

47. Part of the price of borrowing the *Secretum*'s Alexander is the loss of Alexander's voice. Here, indeed, Gower manages to mimic his source most accurately, for the *Secretum* is a one-sided correspondence, and Alexander's virtue resides in his receiving it—not, as in the case of *Alexander and Dindimus*, in responding. His role in book 7 is that of passive recipient of Aristotle's lore; an active Alexander is an Alexander about to get into trouble, open to the conventional *de casibus* moralizing evident in book 3. (The analogy between Alexander and Amans—also essentially voiceless during the lecture of book 7—is also apparent here). Similar difficulties arise later for Gower when he tries to offer Alexander as a specific model for Henry IV in *In Praise of Peace*; see Grady, "The Lancastrian Gower."

48. "Calistre" refers to Callisthenes, believed (mistakenly) in some traditions to have been one of Alexander's teachers; see T.S. Brown, "Callisthenes and Alexander," *American Journal of Philology* 70 (1949): 225–48. Manzalaoui (" 'Noght in the Registre,' " p. 165–66 n.15) and Hamilton ("Studies in the Sources of Gower," pp. 509–11) claim that Gower found the name in Latini's *Tresor*, where "Calistere" appears as a variant; the interpolated versions of the *Historia de Preliis* also include him, however, and his name appears as "Castor" at *Wars* 767.

49. The former observation is that of Minnis, in "John Gower, *Sapiens* in Ethics and Politics," *MAE* 49 (1980): 207–29, here p. 216; the latter remark is Manzalaoui's, from " 'Noght in the Registre,' " p. 165. The last century of Gower criticism has had much to say about the relationship of book 7 to the rest of the poem. Macaulay, in the first *Cambridge History of English Literature*, deemed it "absolutely irrelevant" to the main subject of the *Confessio*, C.S. Lewis considered it a large-scale digression, and generations of readers interested in Gower's exempla have frequently found its matter tedious. For the last several decades, though, critics have (correctly, I think) seen the seventh book as contributing to, rather than detracting from, the *Confessio*'s unity; it has been characterized as central to the poem, "the heart of the discussion," structurally the most important book in the poem, and a recapitulation of the poem's major themes. For Macaulay's opinion see his essay on Gower in the *Cambridge History of English Literature*, vol. 2, *The End of the Middle Ages*, ed. A.W. Ward and A.P. Waller (Cambridge: Cambridge University Press, 1908), esp. pp. 169, 171–72; for Lewis, see *The Allegory of Love* (Oxford: Clarendon Press, 1936), pp. 213–14. The other four opinions belong to George R. Coffman, "John Gower in His Most Significant Role," in *Elizabethan studies and Other Essays in Honor of George F. Reynolds*, University of Colorado Studies, Series B, 2:4 (Boulder, CO, 1945): 52–61;

John H. Fisher, *John Gower: Moral Philosopher and Friend of Chaucer* (New York: New York University Press, 1964), pp. 196–8; Russell A. Peck, *Kingship and Common Profit in Gower's Confessio amantis* (Carbondale, IL: Southern Illinois University Press, 1978), p. 140; and Elizabeth Porter, "Gower's Ethical Microcosm and Political Macrocosm," in A.J. Minnis, ed., *Gower's Confessio amantis: Responses and Reassessments* (Cambridge: D.S. Brewer, 1983), pp. 153–60. For a useful synopsis of critical opinions about book 7, see Peter Nicholson, *An Annotated Index to the Commentary on Gower's Confessio amantis*, Medieval and Renaissance Texts and Studies 62 (Binghamton, NY: Medieval & Renaissance Texts and Studies, 1989), pp. 423–26; more recent summaries can also be found in Yeager, *John Gower's Poetic*, pp. 196–201, 207–16.

That book 7 has a central role now represents canonical opinion; see the new *Cambridge History of Medieval English Literature*, where Winthrop Wetherbee writes that "Book 7, coherent and often impressive in itself, can be seen to have an integrative function," and that it seeks to "systematize the linkage of private and public virtue so central to Gower's project"(p. 604). At the same time, Wetherbee is also careful to acknowledge that while book 7 is integral to the theme of the poem, it is not particularly well integrated into the narrative: it breaks the string of books organized according to particular sins; it interrupts the dialogue of Genius and Amans, who speaks for the first time at line 5408 of its 5438 lines; and of course it requires Genius to go beyond his amatory brief. Compare also Scanlon on the same swing of the critical pendulum: "More recent accounts. . .have frequently provided philosophical justification for Gower's rhetorical disjunctions, rightly arguing, for example, that book 7 is central to moral concerns that run through the poem. Nevertheless, the very justice of these accounts threatens to obscure the rhetorical disjunction they are intended to explain. Book 7 *is* a disruption of the penitential frame, and a rather spectacular one at that. Like the other disruptions, it should not be dismissed, but it should not be explained away either. Any attempt to impose a single, unified vision on the poem belies its actual rhetorical complexity" (Scanlon, *Narrative, Authority, and Power*, p. 214).

50. It's possible that what Amans means here is literally "things strange to the topic of love," that is, a new subject entirely, but that would be an unusual sense for Gower, who typically in the *Confessio* uses the word to mean either "standoffish, distant" (cf. 3. 1136–37, "For whan mi fortune overcasteth / Hire whiel and is to me so strange") or, much more commonly, "foreign, unfamiliar": strange lands, strange names, strange foods, "strange matiere" (1.3092).

51. On Gower's relationship with both of these sources, see Simpson, *Sciences and the Self*, pp. 211–28, and Porter, "Gower's Ethical Microcosm." It is worth noting that the correspondence of these two figures was not entirely one-sided in the medieval imagination; the *Secretum* tradition allegedly preserved Aristotle's letter to Alexander, but Alexander's equally famous *Epistola ad Aristotelem* on the natural and monstrous marvels of India represented a

parallel textual tradition, one that we can also see Gower implicitly rejecting at the end of book 6 of the *Confessio*. This letter, widely known and circulated in Latin and translated into both Old English and Middle English, offers yet another characterization of Alexander—the exotic Alexander or Alexander the *curiosus*—and Gower enacts a sort of banishment of *curiositas* in the stories that conclude book 6. There both the well-travelled Ulysses, consorting with sorceresses and obsessing over prophetic dreams, and the foreign necro-mancer Nectanabus, attracting the attention of Olimpias by virtue of his exotic erotic appeal ("The queene on him hire yhe caste, / And knew that he was strange anon," 6.1864–65), come to wretched ends. "To conne moche thing it helpeth, / Bot of to mochel noman yelpeth" (6.2391–2), says Genius in summary, implicitly drawing a distinction between the transgressive knowledge that characterizes the practice of sorcery in book 6 and the more ordered philosophical pursuits to be described in book 7, wherein "wisdom, hou that evere it stonde, / To him that can it understonde / Doth gret profit in sondri wise" (6.2421–23). Simultaneously, Gower disengages Alexander from the related and thus potentially suspect *Epistola* tradition. Alexander's killing of Nectanabus is therefore doubly (and paradoxically) overdeter-mined; it fulfills the prophecy of Nectanabus, and also opportunely banishes from the poem the kind of perverse knowledge that makes such prophecy possible. On the role of *curiositas* in these tales, see Olsson, *John Gower and the Structures of Conversion*, pp. 182–88, where he writes (p. 182), "The last major tale of [book 6] is linked with the whole of book 7 not merely because each centers on a teacher of Alexander, on Nectanabus and Aristotle respectively, but because these teachers manifest the difference between curiosity and *studiositas* or a 'controlled devotion to learning.' "

    See chapter 3, n20, above, for the unique Middle English version of the *Epistola*; the Latin versions circulated widely, most often in manuscripts con-taining the Julius Valerius *Epitome* of Alexander's life. Early English interest in the letter is attested by the Anglo-Saxon version contained in the *Beowulf* MS; see Stanley Rypins, ed., *Three old English prose texts in Ms. Cotton Vitellius A XV*, EETS o.s. 161 (Oxford: Oxford University Press, 1926; repr. 1987), and Cary, *Medieval Alexander*, pp. 14–16.

52. Technically the references to Constantine and Trajan (7.3137–56) praising pity (but omitting both conversion and salvation) refer to post-Incarnation emperors; moreover a few second recension manuscripts contain some exceptions: *2329–2342 quotes Dante on flattery (but appears in only four MSS), while *3149–60 cites James ii. 13, "Cristes lore," and Cassiodorus, in a passage that reworks *Miroir* 13918–31 (and which appears in only three MSS); see Macaulay's notes at ii.319 and 531, and i.161–2 for the relevant passage of the *Miroir*.

Conclusion   Virtuous Pagans and Virtual Jews

1. Mandeville's virtuous Indians "don to no man otherwise than thei wolde that other men diden to hem" (32.211), while Trajan, in Trevisa's translation

of the *Polychronicon*, vows to "be suche an emperour to oþer men as y wolde þat þey were to me and þey were emperours." See chap. 1 n.5.

2. For this text see Robert Steele, ed., *Three Prose Versions of the Secreta Secretorum*, EETS e.s. 74 (London: Kegan Paul, Trench, Trübner & Co., 1898), pp. 41–118; the story of the Jew and the "enchauntere of þe orient" appears on pp. 104–06. On the origins of the tale, see M.A. Manzalaoui, " 'Noght in the Registre ofVenus': Gower's English Mirror for Princes," in *Medieval Studies for J.A.W. Bennett (Aetatis Suae lxx)*, ed. P.L. Heyworth (Oxford: Clarendon press, 1981), p. 173.

3. Ed. Steele, *Three Prose Versions*, p. 106. For discussion of Gower's alterations see Elizabeth Porter, "Gower's Ethical Microcosm and Political Macrocosm," in *Gower's Confessio amantis : Responses and Reassessments*, ed., A.J. Minnis (Cambridge : D.S. Brewer, 1983), p.155, and Dorothy Metlitzki, *The Matter of Araby in Medieval England* (New Haven:Yale University Press, 1977), p. 110. Also noteworthy is the fact that in Gower, it is the pagan who first articulates his law, not the Jew.

4. See François Hartog, *The Mirror of Herotodus: The Representation of the Other in the Writing of History*, trans. Janet Lloyd (Berkeley and Los Angeles: University of California Press, 1988), pp. 258–59. I owe this citation to Alan S.Ambrisco, who smartly uses "the rule of the excluded middle" to describe the relationship between the audience, the court of Cambyuskan, and the gift-bearing knight in Chaucer's *Squire's Tale* in his " 'It Lyth Nat in My Tonge': Occupatio and Otherness in the *Squire's Tale*," *ChR* 38: 3 (2004): 205–28, esp. 213–14.

5. For another account of the way "[n]arrative eludes maxim" in the *Secretum* version of the Jew–Pagan story, see Judith Ferster, *Fictions of Advice: The Literature and Politics of Counsel in Late Medieval England* (Philadelphia: University of Pennsylvania Press), pp. 52–53.

6. The bibliography on medieval anti-Semitism and Jewish-Christian relations in the Middle Ages more generally is too large and too quickly growing to be adequately captured in a footnote; nevertheless, some key texts (with an English emphasis) are Lester K. Little, *Religious Poverty and the Profit Economy in Medieval Europe* (Ithaca: Cornell University Press, 1978); Jeremy Cohen, *The Friars and the Jews: The Evolution of Medieval Anti-Judaism* (Ithaca: Cornell University Press, 1982); R.I Moore, *The Formation of a Persecuting Society* (Oxford: Basil Blackwell, 1987); Gavin Langmuir, *Toward a Definition of Antisemitism* (Berkeley: University of California Press, 1990); Langmuir, *History, Religion, and Antisemitism* (Berkeley: University of California Press, 1990); Denise L. Despres, "Cultic Anti-Judaism and Chaucer's Litel Clergeon," *Modern Philology* 91 (1994): 413–27; Jeremy Cohen, ed., *From Witness to Witchcraft: Jews and Judaism in Medieval Thought*, Wolfenbutteler Mittelalter-Studien, Band 11(Weisbaden: Harrassowitz Verlag, 1996); Steven Kruger, "The Spectral Jew," *NML* 2 (1998): 9–35; Sylvia Tomasch, "Judecca, Dante's Satan, and the Dis-placed Jew," in *Text and Territory*, ed. Tomasch and Sealy Gilles (Philadelphia: University of Pennsylvania Press, 1998), pp. 247–67; Sylvca Tomasch, "Postcolonial Chaucer and the Virtual Jew," in

The Postcolonial Middle Ages, ed. Jeffrey Jerome Cohen (New York: St. Martin's Press, 2000), pp. 243–60; Elisa Narin van Court, "Socially Marginal, Culturally Central: Representing Jews in Lated Medieval English Literature," *Exemplaria* 12 (2000): 293–326; Lee Patterson, " 'The Living Witness of Our Redemption': Martyrdom and Imitation in Chaucer's *Prioresse's Tale*," *JMEMS* 31:3 (2001): 507–60; Sheila Delany, ed., *Chaucer and the Jews: Sources, Contexts, Meanings* (New York: Routledge, 2002); Lisa Lampert, *Gender and Jewish Difference from Paul to Shakespeare* (Philadelphia : University of Pennsylvania Press, 2004).

7. On this passage see Benjamin Braude, "*Mandeville's* Jews Among Others," in *Pilgrims and Travellers to the Holy Land*, ed. Bryan F. LeBeau and Menachem Mor, Studies in Jewish Civilization 7 (Omaha, NE: Creighton University Press, 1995), pp. 145–49, and Iain Higgins, *Writing East: The "Travels" of Sir John Mandeville* (Philadelphia: University of Pennsylvania Press, 1997), pp. 178–89.

8. "And other trees that beren venym ayenst the whiche there is no medicyne but [on], and that is to taken here propre leves and stampe hem and tempere hem with water and than drynke it; and elles he schalle dye, for triacle wil not avaylle ne non other medicyne. Of this venym the Iewes had let seche of on of here frendes for to enpoysone alle Cristiantee, as I haue herd hem seye in here confessioun before here dyenge. But, thanked be alemyghty God, thei fayleden of hire purpos, but alleweys thei maken gret mortalitee of people" (21.139–40). The accusation of course echoes contemporary charges that the Jews caused the plague by poisoning wells. The passage, typically Mandevillean in its assumption that "confessioun" is part of Jewish practice, is likely an interpolation into the Cotton text, as it is absent in the earlier Defective Version; see ed. M.C. Seymour, *The Defective Version of Mandeville's Travels*, EETS o.s. 319 (Oxford: Oxford University Press, 2002), p. 83.

9. See chap. 2 n.4 above.

10. Stephen Greenblatt, *Marvelous Possessions: The Wonder of the New World* (Chicago: University of Chicago Press, 1991), p. 51. David Lawton, "The Surveying Subject and the 'Whole World' of Belief: Three Case Studies," *NML* 4 (2001): 25, makes the provocative observation that in his itinerary, "Sir John" resembles the homeless Jew: "Taken seriously, he is ubiquitous and homeless; he, not the Sultan or the Khan, is the 'other' of the text, at home in every language, the eternal foreigner: the 'perennial other,' 'radically heteronomous.' I am citing Levinas here, who goes on to add: 'Jew'. . .It is *Mandeville's* aporia; the text's kinship with what it most hates."

11. For the term "virtual Jew" see Tomasch, "Postcolonial Chaucer and the Virtual Jew," esp. pp. 252–54. "Hermeneutical Jew" is the coinage of Jeremy Cohen; see his "Introduction" in *From Witness to Witchcraft*, p. 9.

12. From Bodleian Lib. MS Rawl. B.490, ed. Steele, *Three Prose Versions*, p. 201. On Yonge—who tells the story under the rubric "That god nath not in dispite the orisones of Paganes"—see *DNB* XXI.1240–41. Yonge also tells the Jew/Pagan story in his translation; he makes the story serve under the rubric of the cardinal virtues, specifically justice, and introduces it with

citations from Bernard and Solomon to the effect "That a prynce sholde not truste to his enemy"—admonitions immediately applied to the Anglo-Irish milieu of *The Gouernaunce of Prynces*, a text dedicated to James Butler, Earl of Ormond and Henry V's Irish lieutenant at the time: apparently the Irish are as little to be trusted as the Jews. Ferster, *Fictions of Advice*, pp. 55–66, discusses the context and likely meaning of Yonge's dedication.

Interestingly Yonge also tells, in his very next chapter, the Gregory/Trajan story, completing the virtuous pagan trifecta, and interestingly he manages to put a unique anti-Jewish slant on that tale, too. In between his accounts of Trajan's exemplary justice (attributed to Helinandus) and Gregory's intercessory prayers, he elaborates on the applicability of Trajan's example to Christian lords: "Moche sholde oure crystyn Prynces reede and be ashamyd, whan thay doth no ryght to the Pepill, or slackely and Slowely hare wrongis amendyth, whan Iusticia, as well to Pouer as to ryche sholde be done frely, Delayeth for fawoure or for hate, or hit for Penyes sylle and Sauyth gilti men, and dampnyth gylteles men. Tho men ben lykenyd to the Iues, the cruel fellons, the whyche Sauyd baraban the thefe and a man murderere, and crucifieddyn Ihesu, the verray Sauyoure" (ed. Steele, pp. 168–69). Be like Trajan, or be like the Jews—that's the simple choice Yonge offers, establishing in the starkest terms possible the structural relationship between pagan virtue and Jewish vice.

The introduction of the distinction between poor and rich here interestingly echoes Langland's account of Trajan, whose appearance is immediately followed by a long defense and praise of poverty at B.11.170ff; though the critique of unequal justice was certainly a commonplace in the period, might it be worth thinking about the possible influence of *Piers Plowman* here and elsewhere in Yonge's considerably amplified translation?

13. In the very next line, the Saracens also condemn the Christians, a juxtaposition permits us to see abjection in action, as it were: "And the Cristene ben cursed also, as thei seyn, for thei kepen not the commandementes and the preceptes of the gospelle that Ihesu Crist taughte hem."

14. The distinction between potentially righteous Hebrews and cursed Jews is a staple of medieval Christian writing about Jews; see e.g, Kruger, "Spectral Jew," pp.12–15, and Narin van Court, "Socially Marginal," pp. 298–308. As Sylvia Tomasch has shown, the binary also informs Dante's *Commedia*, another poem very interested in virtuous pagans; see her "Judecca, Dante's Satan, and the *Dis*-placed Jew," in *Text and Territory*, pp. 247–67. Ironically enough, in post-Expulsion England the most likely encounter for an inhabitant of the London of Langland or Gower to have with living Jews would have been with a converted Jew; on the history of the Domus Conversorum, the establishment that housed such individuals, see Henry Ansgar Kelly, "Jews and Saracens in Chaucer's England," *SAC* 27 (2005): 125–65.

15. Mandeville's version appears in Chapter 11 of the Cotton text.

16. The definitive account of the poem's manuscripts, sources, and origin is the edition of Ralph Hanna and David Lawton, *The Siege of Jerusalem*, EETS

o.s. 320 (Oxford: Oxford University Press, 2003). All quotations are drawn from this edition. Notable among the surviving MSS are two in which the poem appears with *Piers Plowman* (the C-version in Bodleian Lib. MS Laud Misc. 656, the B-version in Huntington Lib. MS HM 128) and one (Brit. Lib. MS Add. 31042) copied by Robert Thornton, who also copied the Alliterative *Morte Arthure*. On the last see John J. Thompson, *Robert Thornton and the London Thornton Manuscript: British Library MS Additional 31042* (Cambridge: D.S. Brewer, 1987).

17. The poet's lurid attention to the grisly details of Jewish suffering during the siege essentially put the poem off-limits to modern criticism until relatively recently; in fact such an observation has itself become commonplace of the growing body of fine essays and chapters devoted to the *Siege*, which includes Ralph III Hanna, "Contextualizing *The Siege of Jerusalem*," *YLS* 6 (1992): 109–21; Elisa Narin van Court, "*The Siege of Jerusalem* and Augustinian Historians: Writing About Jews in Fourteenth-Century England," *ChR* 29 (1995): 227–48; David Lawton, "Titus Goes Hunting and Hawking: The Poetics of Recreation and Revenge in *The Siege of Jerusalem*," in *Individuality and Achievement in Middle English Poetry*, ed. O.S. Pickering (Cambridge: D.S. Brewer, 1997), pp. 105–117; Christine Chism, *Alliterative Revivals* (Philadelphia: University of Pennsylvania Press, 2002), pp. 155–88; and Roger Nicholson, "Haunted Itineraries: Reading *The Siege of Jerusalem*," *Exemplaria* 14 (2002): 447–84. There is also a book-length study: Bonnie Millar, *The Siege of Jerusalem in its Physical, Literary, and Historical Contexts* (Dublin: Four Courts Press, 2000).

18. Hanna "Contextualizing," p. 113.

19. Chism, *Alliterative Revivals*, p. 181.

20. The phrase is Hanna's, from "Contextualizing," p. 112. Suzanne M. Yeager, in "*The Siege of Jerusalem* and Biblical Exegesis: Writing about Romans in Fourteenth-Century England," *Chaucer Review* 39.1 (2004): 70–102, describes the double valence of Rome both in the poem and in contemporary exegetical writing; surveying the literature, she observes, "one finds that [the Romans] are depicted as depraved persecutors of the faithful *and* as victorious warriors for Christ" (p. 71).

21 Kathleen Biddick, *The Shock of Medievalism* (Durham and London: Duke University Press, 1998), p. 14.

22. The one exception to this rule is the Jewish historian Josephus, who accompanies Titus back to Rome, "þer of þis mater and mo he made fayre bokes" (1326). As Chism observes, "if a Jewish figure can be coopted to agree to his own supersession, Josephus is he" (p. 184).

On the relevance of contemporary crusading to the *Siege*, see Chism, *Alliterative Revivals*, p. 169, who observes astutely that "*The Siege of Jerusalem* shows how the Muslim supersession of Christianity darkens medieval Christianity's views of its own Jewish forefather. The poem reflects the troubled consciousness of an Augustinian Christianity caught in its own supersessional dialectic and, unable to dislodge its successor from the Holy Land, taking 'great consolation' in turning its frustration against a more vulnerable precursor." See also Nicholson, "Haunted Itineraries"; Millar, *Siege*, pp. 147ff; and Mary Hamel, "*The Siege of Jerusalem* as a Crusading Poem," in *Journeys to God: Pilgrimage and Crusade*, ed. Barbara N. Sargent-Baur (Kalamazoo, MI: Medieval Institute Publications, 1992), pp. 177–94.

23. "A Week in the Life of a High School," *Time Magazine*, October 25, 1999, p. 110.

BIBLIOGRAPHY

Abbreviations

CBMLC	Corpus of British Medieval Library Catalogues
ChR	*Chaucer Review*
EETS e.s.	Early English Text Society, extra series
EETS o.s.	Early English Text Society, original series
EETS s.s.	Early English Text Society, supplementary series
JMEMS	*Journal of Medieval and Modern Studies*
MAE	*Medium Ævum*
NML	*New Medieval Literatures*
SAC	*Studies in the Age of Chaucer*
YLS	*Yearbook of Langland Studies*

Manuscripts

Cambridge, Cambridge University Library MS Ll.I.15
Cambridge, Corpus Christi College, Cambridge MS 219
Cambridge, Corpus Christi College, Cambridge MS 450
Cambridge, Gonville and Caius College MS 154
Cambridge, Gonville and Caius College MS 230/116
Glasgow, Glasgow University Library MS Hunter 84
London, British Library MS Arundel 123
Oxford, Bodleian Library MS Eng. Poet. a. 1 (Vernon MS)
Oxford, Bodleian Library MS Bodley 264

Primary Texts

Alexander and Dindimus. Ed. W.W. Skeat. EETS o.s.31. London: H. Milford for Oxford University Press, 1878; repr.1930.
Alexandri Magni Iter ad Paradisu. Ed. Mario Esposito. *Hermathena* XV (1909): 368–83.
Alexandri Magni Iter ad Paradisum. Ed. Julius Zacher. Konigsberg: T. Theile, 1859.
An Alphabet of Tales. Ed. Mary Macleod Banks. EETS o.s. 126–27. London: Kegan Paul, Trench, Trübner and Co., 1904–05.

Der altfranzösiche prosa-Alexanderroman nach der Berliner bilderhandschrift nebst dem lateinischen original der Historia de Preliis. Ed. Alfons Hilka. Halle a S.: M. Niemeyer, 1920.

Aquinas, Thomas. *Opera Omnia*. 25 vols. Parma: Petri Fiaccadori, 1852–73; repr. NY: Musurgia Publishers, 1948.

———. *Summa Theologica*. Trans. Fathers of the English Dominican Province. 22 vols. London: Burns, Oates and Washbourne, Ltd., 1921–24.

Augustine. *City of God*. Trans. Henry Bettenson. New York: Penguin, 1980.

Bacon, Roger. *Opera hactenus inedita Rogeri Baconi*. Fasc. 5. Oxford: Clarendon Press, 1920.

The Book of Vices and Virtues. Ed. W. Nelson Francis. EETS o.s. 217. Oxford: Oxford University Press, 1942; repr. 1968.

Boccaccio, Giovanni. *The Book of Theseus: Teseida delle Nozze d'Emilia*. Trans. Bernadette Marie McCoy. New York: Medieval Text Association, 1974.

Caesarius of Heisterbach. *Dialogus Miraculorum*. Ed. J. Strange. Cologne: H. Lempertz & Co., 1851.

———. *The Dialogue on Miracles*. Ed. and trans. E. von E. Scott and C.C. Swinton Bland. London: G. Routledge & sons, ltd., 1929.

Capgrave, John. *Abbreuiacion of Cronicles*. Ed. Peter J. Lucas. EETS o.s. 285. Oxford: Oxford University Press, 1983.

Chappell, Julie. "The Prose 'Alexander' of Robert Thornton: The Middle English Text with a Modern English Translation." Ph.D. Diss, University of Washington, 1982.

Chaucer, Geoffrey. *The Book of Troilus and Criseyde*. Ed. R.K. Root. Princeton: Princeton University Press, 1926; repr. 1945.

———. *The Riverside Chaucer*. Ed. Larry D. Benson. 3rd edn. Boston: Houghton Mifflin, 1987.

Chaucer Life-Records. Ed. Martin M. Crow and Clair C. Olson. Oxford: Clarendon Press, 1966.

Chronica Monasterii de Melsa. Ed. E.A. Bond. 3 vols. Rolls Series 43. London: Longmans, Green, Reader, and Dyer, 1866–68.

Dan Michel's Ayenbite of Inwit. Ed. Pamela Gradon. EETS o.s. 278. Oxford: Oxford University Press, 1979.

Dante. *The Divine Comedy*. Trans. Charles S. Singleton. Princeton: Princeton University Press, 1973.

Dawson, Christopher. *Mission to Asia*. Medieval Academy Reprints 8. Toronto: University of Toronto Press, 1980.

The Defective Version of Mandeville's Travels. Ed. M.C. Seymour. EETS o.s. 319. Oxford: Oxford University Press, 2002.

Dimarco, Vincent, and Leslie Perelman, eds. "The Middle English Letter of Alexander to Aristotle." *Costerus* 13 (1978).

The Earliest Life of Gregory the Great. Ed. and trans. Bertram Colgrave. Lawrence, KS: University of Kansas Press, 1968.

Eulogium Historiarum sive Temporis. Ed. F.S. Haydon. 3 vols. Rolls Series 9. London: Longman, Brown, Green, Longmans and Roberts, 1858–63.

The Gests of King Alexander of Macedon. Ed. F.P. Magoun. Cambridge, MA: Harvard University Press, 1929.

Given-Wilson, Chris, ed. *Chronicles of the Revolution, 1397–1400: The Reign of Richard II.* Manchester: Manchester University Press, 1993.

Gower, John. *Complete Works.* Ed. G.C. Macaulay. 4 vols. Oxford: Clarendon Press, 1899–1902.

Gregory the Great. *The Letters of Gregory the Great.* Ed. and trans. John R.C. Martyn. 3 vols. Mediaeval Sources in Translation 40. Toronto: Pontifical Institute of Mediaeval Studies, 2004.

Hahn, Thomas. "The Middle English *Letter of Alexander to Aristotle*: Introduction, Text, Sources, and Commentary." *Medieval Studies* XLI (1979): 106–45.

Hayton. *Fleurs des Histors D'Orient.* In *Recueil des Historiens des Croisades: Documentes Armeniens.* Ed. Éd. Dulaurier. 2 vols. Paris: Imprimerie Nationale, 1869–1906.

———. *A Lytell Cronycle: Richard Pynson's Translation (ca. 1520) of La Fleur des histoires de la terre d'Orient (ca. 1307).* Ed. Glenn Burger. Toronto Medieval Texts and Translations 6. Toronto: University of Toronto Press, 1988.

Higden, Ranulph. *Polychronicon Ranulphi Higden monachi Cestrensis.* Ed. J.R. Lumby. 9 vols. Rolls Series 41. London: Longman & Co., 1865–86.

Die Historia de Preliis Alexandri Magni—Synoptische Edition der Rezensionen des Leo Archipresbyter und die interpolierten Fassungen J¹, J², J³ (Buch I & II). Ed. Hermann-Josef Bergmeister. *Beiträge zur klassischen Philologie,* Heft 65. Meisenheim am Glan: A. Hain, 1975.

Die Historia de preliis Alexandri Magni: Rezension J3. Ed. Karl Steffens. *Beiträge zur klassischen Philologie,* Heft 73. Meisenheim am Glan: A. Hain, 1975.

Horstmann, Carl. "Die Evangelien-Geschichten der Homiliensammlung des MS. Vernon." *Archiv für das Studium der neueren Sprachen und Literaturen* 57 (1877): 241–316.

Hulton, W.A., ed. *Documents Relating to the Priory of Penwortham, and other Possessions in Lancashire of the Abbey of Evesham.* Manchester: Chetham Society, 1853.

Isidore of Seville. *Chronica Maiora Isidori Iunioris.* Ed. T. Mommsen. *Monumenta Germaniae Historica, Auctores Antiqquissimorum* XI. 2 vols. Berlin: Weidman, 1884.

Jacobus de Voragine. *Legenda Aurea Vulgo Historia Lombardica Dicta.* Ed. Th. Graesse. 2nd edn. Leipzig: Librariae Arnoldianae, 1850.

———. *The Golden Legend.* Trans. William Granger Ryan. 2 vols. Princeton: Princeton University Press, 1995.

Lambert le Tort. *The Romance of Alexander: A Collotype Facsimile of MS Bodley 264.* Introd. M.R. James. Oxford: Clarendon, 1933.

Langand, William. *Piers Plowman: The B Version.* Ed. George Kane and E. Talbot Donaldson. London: Athlone Press, 1975.

———. *Piers Plowman: An Edition of the C-Text.* Ed. Derek Pearsall. Berkeley: University of California Press, 1978.

Lydgate, John and Benedict Burgh. *Secrees of Old Philosoffres.* Ed. Robert Steele. EETS o.s. 66. London: K. Paul, Trench, Trübner, 1894.

Malory, Sir Thomas. *Works.* Ed. Eugène Vinaver. 2nd. edn. Oxford: Oxford University Press, 1977.

Mandeville's Travels. Ed. P. Hamelius. 2 vols. EETS o.s. 153 & 154. London: Kegan Paul, 1919.

Mandeville's Travels. Ed. M.C. Seymour. Oxford: Oxford University Press, 1967.

Mandeville's Travels: Texts and Translations. Ed. Malcolm Letts. Hakluyt Society, 2nd ser. 101–102. London: Hakluyt Society, 1953.

Morte Arthure. Ed. John Finlayson. Evanston, IL: Northwestern University Press, 1967.

Morte Arthure: A Critical Edition. Ed. Mary Hamel. New York: Garland Publishing, 1984.

Neeson, Marjorie. "*The Prose Alexander: A Critical Edition*." Ph.D. Diss., UCLA, 1971.

Nevanlinna, Saara, ed. *The Northern Homily Cycle: The Expanded Version in MSS Harley 4196 and Cotton Tiberius E. VII*. Memoires de la Société Néophilologique de Helsinki 38 (1972).

Philippe de Mézières. *Le Songe du Vieil Pelerin*. Ed. G.W. Coopland. 2 vols. Cambridge: Cambridge University Press, 1969.

———. *Letter to King Richard II*. Ed. G.W. Coopland. Liverpool: Liverpool University Press, 1975.

The Prose Life of Alexander. Ed. J.S. Westlake EETS o.s. 143. London: Kegan Paul, Trench, Trübner and Co., 1913.

St. Erkenwald. Ed. Sir Israel Gollancz. Oxford: Oxford University Press, 1922.

St. Erkenwald. Ed. Henry L. Savage. Yale Studies in English 72. New Haven: Yale University Press, 1926.

St. Erkenwald. Ed. Ruth Morse. Cambridge: D.S. Brewer, 1975.

St. Erkenwald. Ed. Clifford Peterson. Philadelphia: University of Pennsylvania Press, 1977.

Schlauch, Margaret. *Medieval Narrative: A Book of Translations*. New York: Prentice-Hall, Inc., 1928.

Secretum Secretorum: Nine English Versions. Ed. M.A. Manzalaoui. EETS o.s. 276. Oxford: Oxford University Press, 1977.

The Siege of Jerusalem. Ed. Ralph Hanna and David Lawton. EETS o.s. 320. Oxford: Oxford University Press, 2003.

Sir Ferumbras. Ed. Sidney J. Herrtage. EETS o.s. 34. London: Kegan Paul, Trench Trübner & Co., 1879.

Testamenta Eboracensa. Ed. J. Raine. Surtees Society 30 (1855).

Three Old English Prose Texts in Ms. Cotton Vitellius A XV. Ed. Stanley Rypins. EETS o.s. 161. Oxford: Oxford University Press, 1926; repr. 1987.

Three Prose Versions of the Secreta Secretorum. Ed. Robert Steele. EETS e.s. 74. London: K. Paul, Trench, Trübner, 1908.

The Vernon Manuscript: A Facsimile of Bodleian Library, Oxford, MS Eng. Poet. a. 1. Intr. A.I. Doyle. Cambridge: D.S. Brewer, 1987.

Vincent of Beauvais. *Bibliotheca Mundi seu Speculum Maioris Vincenti Bvrgundi Praesvlis Bellovacensis*. 4 vols. Douai, 1624; repr. Graz: Akademische Druck-u. Verlaganstalt, 1964–65.

Walsingham, Thomas. *Johannis de Trokelowe er Henrici de Blaneforde. . .chronica et annales*. Ed. H.T. Riley. Rolls Series 28:4. London: Longmans, Green, Reader and Dyer, 1866.

———. *Annales Monasterii S. Albani*. Ed. H.T. Riley. 2 vols. Rolls Series 28. London: Longmans, Green, 1870–71.

————. *Ypodigma Neustriae*. Ed. H.T. Riley. Rolls Series 28:7. London: Longman, 1876.

————. *The St. Albans Chronicle 1406–1420*. Ed. V.H. Galbraith. Oxford: Clarendon Press, 1937.

The Wars of Alexander. Ed. Hoyt N. Duggan and Thorlac Turville-Petre. EETS s.s. 10. Oxford: Oxford University Press, 1989.

William of Tripoli. *Tractatus de statu Saracenorum*. Ed. Hans Prutz. *Kulturgeschichte der Kreuzzuge* (1883): 573–98.

Wynnere and Wastoure. Ed. Stephanie Trigg. EETS o.s. 297. Oxford: Oxford University Press, 1990.

Secondary Sources

"A Week in the Life of a High School." *Time Magazine*. October 25, 1999. 66–115.

Adams, Robert. "Piers's Pardon and Langland's Semi-Pelagianism." *Traditio* 39 (1983): 367–418.

————. "Langland's Theology." In *A Companion to Piers Plowman*. Ed. John A. Alford. Berkeley: University of California Press, 1988. 87–114.

Aerts, W.J., E.R. Smits and J.B. Voorbij, eds. *Vincent of Beauvais and Alexander the Great: Studies on the Speculum Maius and its translations into Medieval vernaculars*. Medievalia Groningana fasc. 7. Groningen: E. Forsten, 1986.

Alford, John A., ed. *A Companion to Piers Plowman*. Berkeley: University of California Press, 1988.

Altman, Janet. *Epistolarity: Approaches to a Form*. Columbus, OH: The Ohio State University Press, 1982.

Ambrisco, Alan S. " 'It Lyth Nat in My Tonge': Occupatio and Otherness in the *Squire's Tale*." *ChR* 38:3 (2004): 205–28.

Anderson, Andrew Runni. *Alexander's Gate, Gog and Magog, and the Inclosed Nations*. Cambridge, MA: Medieval Academy of America, 1932.

Attiya, Hussein M. "Knowledge of Arabic in the Crusader States in the twelfth and thirteenth centuries." *Journal of Medieval History* 25 (1999): 203–13.

Baker, Denise. "From Plowing to Penitence: *Piers Plowman* and Fourteenth-Century Theology." *Speculum* 55 (1980): 715–25.

Bankert, Dabney Anderson. "Secularizing the Word: Conversion Models in Chaucer's *Troilus and Criseyde*." *ChR* 37 (2003): 196–218.

Becker, H. "Zur Alexandersage." *Zeitschrift für Deutsche Philologie* 23 (1891): 424–25.

Bell, David N. ed. *The Libraries of the Cistercians, Gilbertines, and Premonstratensians*. CBMLC 3. London: The British Library, 1992.

Bennett, J.A.W. "Chaucer's Contemporary." In *Piers Plowman: Critical Approaches*. Ed. S.S. Hussey. London: Methuen, 1969.

Bennett, J.A.W. *Middle English Literature*. Ed. and compl. by Douglas Gray. Oxford: Clarendon, 1986.

Bennett, J.W. *The Rediscovery of Sir John Mandeville*. New York: Modern Language Association of America, 1954.

Benson, Larry. "The Alliterative *Morte Arthure* and Medieval Tragedy." *Tennessee Studies in Literature* 11 (1966): 75–86.

Biddick, Kathleen. "The ABC of Ptolemy: Mapping the World with the Alphabet." In *Text and Territory: Geographical Imagination in the European Middle Ages*. Ed. Sylvia Tomasch and Sealy Gilles. Philadelphia: University of Pennsylvania Press, 1988. 268–93.

———. *The Shock of Medievalism*. Durham and London: Duke University Press, 1998.

Blaess, Madeleine. "L'Abbaye de Bordesley et les Livres de Guy de Beauchamp." *Romania* LXXVIII (1957): 511–18.

Braude, Benjamin. "*Mandeville's* Jews Among Others." In *Pilgrims and Travellers to the Holy Land*. Ed. Bryan F. LeBeau and Menachem Mor. Studies in Jewish Civilization 7. Omaha, NE: Creighton University Press, 1995. 141–68.

———. "The Sons of Noah and the Construction of Ethnic and Geographical Identities in the Medieval and Early Modern Periods." *The William and Mary Quarterly* 54 (1997): 103–42.

The British Library Catalogue of the Additions to the Manuscripts 1951–5. The British Library, 1982.

The British Library Catalogue of the Additions to the Manuscripts: The Yelverton Manuscripts. 2 vols. London: The British Library, 1994.

Brown, T.S. "Callisthenes and Alexander." *American Journal of Philology* 70 (1949): 225–48.

Bunt, G.H.V. "Alexander and the Universal chronicle: Scholars and Translators." In *The Medieval Alexander Legend and Romance Epic: Essays in Honour of David J.A. Ross*. Ed. Peter Noble, Lucie Polak, and Claire Isoz. Millwood, NJ: Kraus International Publications, 1982. 1–10.

———. "The Story of Alexander the Great in the Middle English translations of Higden's *Polychronicon*." In *Vincent of Beauvais and Alexander the Great: Studies on the Speculum Maius and Its Translations into Medieval Vernaculars*. Ed. W.J. Aerts, E.R. Smits and J.B. Voorbij. Mediaevalia Groningana fasc. 8. Groningen: E. Forsten, 1986. 127–40.

———. "Exemplum and Tale in John Gower's 'Confessio amantis'." In *Exemplum et Similitudo: Alexander the Great and other heroes as points of reference in medieval literature*. Ed. W.J. Aerts and M. Gosman. Mediaevalia Groningana fasc. 8. Groningen: E. Forsten, 1988. 145–57.

———. "The Art of a Medieval Translator: The Thornton Prose Life of Alexander." *Neophilologus* 76 (1992): 147–59.

Burnley, J.D. *Chaucer's Language and the Philosophers' Tradition*. Cambridge: D.S. Brewer, 1979.

Butturf, Douglas R. "Satire in *Mandeville's Travels*." *Annuale Medievale* 13 (1972): 155–64.

Campbell, Mary B., *The Witness and the Other World: Exotic European Travel Writing 400–1600*. Ithaca, NY: Cornell University Press, 1988.

Capéran, Louis. *Le problème du salut des infidèles: essai historique*. 2nd. edn. Toulouse: Grand Seminaire, 1934.

Cary, George. "A Note on the Mediaeval History of the *Collatio Alexandri cum Dindimo*," *Classica et Mediaevalia* XV (1954): 124–29.

———. *The Medieval Alexander*. Ed. D.J.A Ross. Cambridge: Cambridge University Press, 1956; repr. 1976.

A Catalogue of the Manuscripts Preserved in the Library of the University of Cambridge. 6 vols. 1854; repr. Munich: Kraus Reprint, 1980.

Chambers, R.W. "Long Will, Dante, and the Righteous Heathen." *Essays and Studies* IX (1923): 50–69.

———. *Man's Unconquerable Mind*. London: J. Cape, 1939.

Chauvin, Victor. "Le Prétendu Séjour de Mandeville en Égypte." *Wallonia* X (1902): 237–42.

Cheney, C.R. *The English Church and its Laws, 12th–14th Centuries*. London: Variorum, 1982.

Chism, Christine. *Alliterative Revivals*. Philadelphia: University of Pennsylvania Press, 2002.

Clark, James G. "Thomas Walsingham Reconsidered: Books and Learning at Late-Medieval St. Albans." *Speculum* 77 (2002): 832–60.

Coffman, George R. "John Gower in His Most Significant Role." *Elizabethan Studies and Other Essays in Honor of George F. Reynolds*. University of Colorado Studies, Series B, 2:4 (Boulder, CO, 1945): 52–61.

Cohen, Jeffrey Jerome. *Of Giants: Sex, Monsters, and the Middle Ages*. Minneapolis: University of Minnesota Press, 1999.

———. "On Saracen Enjoyment: Some Fantasies of Race in Late Medieval France and England." *JMEMS* 31 (2001): 124–34.

Cohen, Jeremy. *The Friars and the Jews: The Evolution of Medieval Anti-Judaism*. Ithaca, NY: Cornell University Press, 1982.

———, ed. *From Witness to Witchcraft: Jews and Judaism in Medieval Thought*. Wolfenbutteler Mittelalter-Studien, Band 11. Weisbaden: Harrassowitz Verlag, 1996.

Coleman, Janet. *Piers Plowman and the Moderni*. Rome: Edizioni di storia e letteratura, 1981.

Coletti, Theresa. *Naming the Rose: Eco, Medieval Signs, and Modern Theory*. Ithaca, NY: Cornell Unversity Press, 1988.

Colish, Marcia L. "The Virtuous Pagan: Dante and the Christian Tradition." In *The Unbounded Community: Papers in Christian Ecumenism in Honor of Jaroslav Pelikan*. Ed. William Caferro and Duncan G. Fisher. New York: Garland, 1996. 43–92.

Daniel, Norman. *The Arabs and Medieval Europe*. London: Longman, 1975.

Delany, Sheila, ed. *Chaucer and the Jews: Sources, Contexts, Meanings*. New York: Routledge, 2002.

Deluz, Christiane. *Le livre de Jehan de Mandeville: Une "geographie" au XIVe siecle*. Textes, Etudes, Congres 8. Louvain-la-Neuve: Institut d'études médiévales de l'Université catholique de Louvain, 1988.

Derrett, J.D.M. "The History of Palladius on the Races of India and the Brahmins." *Classica et Mediaevalia* 21 (1960): 64–99.

Despres, Denise L. "Cultic Anti-Judaism and Chaucer's Litel Clergeon." *Modern Philology* 91 (1994): 413–27.

Dillon, Viscount and W. St. John Hope, "Inventory of the goods and chattels belonging to Thomas, Duke of Gloucester. . ." *The Archaeological Journal* 54 (1897): 275–308.

Dinshaw, Carolyn. *Chaucer's Sexual Poetics.* Madison: University of Wisconsin Press, 1989.

Donaldson. E.T. *Speaking of Chaucer.* New York: W.W. Norton, 1970.

Doxsee, Elizabeth. " 'Trew Treuth' and Canon Law: The Orthodoxy of Trajan's Salvation in *Piers Plowman* C-Text." *Neuphilologische mitteilungen* 89 (1988): 295–311.

Doyle, A.I. "An Unrecognized Piece of *Piers the Ploughman's Creed* and Other Work by Its Scribe." *Speculum* 34 (1959): 428–36.

————, and M.B. Parkes. "The Production of Copies of the *Canterbury Tales* and the *Confessio amantis* in the Early Fifteenth Century." In *Medieval Scribes, Manuscripts and Libraries: Essays Presented to N.R. Ker.* Ed. M.B. Parkes and Andrew G. Watson. London: Scolar Press, 1978. 163–210.

Duggan, Hoyt. "The Source of the Middle English *The Wars of Alexander*." *Speculum* 51 (1976): 624–36.

Dunning, T.P. "Langland and the Salvation of the Heathen." *MAE* (1943): 45–54.

————. "God and Man in *Troilus and Criseyde*." In *English and Medieval Studies Presented to J.R.R. Tolkien on the Occasion of His Seventieth Birthday.* Ed. Norman Davis and C.L. Wrenn. London: George Allen & Unwin Ltd., 1962. 164–82.

————. "Action and Contemplation in *Piers Plowman*." In *Piers Plowman: Critical Approaches.* Ed. S.S. Hussey. London: Methuen, 1969. 213–25.

Ferster, Judith. *Fictions of Advice: The Literature and Politics of Counsel in Late Medieval England.* Philadelphia: University of Pennsylvania Press, 1996.

Fineman, Joel. "The History of the Anecdote: Fiction and Fiction." In *The New Historicism.* Ed. H. Aram Veeser. London: Routledge, 1989. 49–76.

Finlayson, John. "Arthur and the Giant of St. Michael's Mount." *MAE* 33 (1964): 112–20.

————. "The Alliterative *Morte Arthure* and *Sir Ferumbras*." *Anglia* 92 (1974): 380–86.

Fisher, John H. *John Gower: Moral Philosopher and Friend of Chaucer.* New York: New York University Press, 1964.

Fleming, John. *Classical Imitation and Interpretation in Chaucer's Troilus.* Lincoln: University of Nebraska Press, 1990.

Foster, Kenelm. *The Two Dantes and Other Studies.* London: Darton, Longman and Todd, 1977.

Foucault, Michel. "What is an author?" In *Contemporary Literary Criticism: Literary and Cultural Studies.* 4th edn. Ed. Robert con Davis and Ronald Schleifer. New York: Longman, 1998. 364–76.

Fowler, Elizabeth. "Chaucer's Hard Cases." In *Medieval Crime and Social Control.* Ed. Barbara Hanawalt and David Wallace. Minneapolis: University of Minnesota Press, 1999. 124–42.

Fradenburg, L.O. Aranye. " 'So that we may speak of them': Enjoying the Middle Ages." *New Literary History* 28 (1997): 205–30.

————. *Sacrifice Your Love: Psychoanalysis, Historicism, Chaucer.* Minneapolis: University of Minnesota Press, 2002.

Gallacher, Patrick J. *Love, the Word and Mercury: A Reading of John Gower's Confessio amantis.* Albuquerque: University of New Mexico Press, 1975.

Galloway, Andrew. "The Rhetoric of Riddling in Late-Medieval England: The 'Oxford' Riddles, the *Secretum philosophorum*, and the Riddles in *Piers Plowman.*" *Speculum* 70 (1995): 68–105.

Gaylord, Alan. "The Lesson of the *Troilus*: Chastisement and Correction." In *Essays on Troilus and Criseyde.* Ed. Mary Salu. Cambridge: D.S. Brewer, 1979. 23–42.

Gilbert, A.H. "Notes on the Influence of the *Secretum Secretorum.*" *Speculum* 3 (1928): 84–98.

Gradon, Pamela. "*Trajanus Redivivus*: Another Look at Trajan in *Piers Plowman.*" In *Middle English Studies Presented to Norman Davis in Honour of His Seventieth Birthday.* Ed. Douglas Gray and E.G. Stanley. Oxford: Clarendon Press, 1983. 93–114.

Grady, Frank. "The Literary and Political Recuperation of Pagan Virtue in the English Middle Ages," Ph.D. diss. U.C. Berkeley, 1991.

———. "*Piers Plowman, St. Erkenwald*, and the Rule of Exceptional Salvations." *YLS* 6 (1992): 61–86.

———. "Machomete and *Mandeville's Travels.*" In *Medieval Christian Perceptions of Islam.* Ed. John Tolan. New York: Garland Publishing, 1994; repr. Routledge, 2002. 271–88.

———. "The Lancastrian Gower and the Limits of Exemplarity." *Speculum* 70 (1995): 552–75.

———. "Chaucer Reading Langland: *The House of Fame.*" *SAC* 18 (1996): 3–23.

———. "*St. Erkenwald* and the Merciless Parliament." *SAC* 22 (2000): 179–211.

———. "Contextualizing *Alexander and Dindimus.*" *YLS* 18 (2004): 81–106.

Green, Peter. *Alexander of Macedon, 356–323 B.C.: A Historical Biography.* Berkeley: University of California Press, 1991.

Green, R.F. *A Crisis of Truth: Literature and Law in Ricardian England.* Philadelphia: University of Pennsylvania Press, 1999.

Greenblatt, Stephen. *Shakespearean Negotiations: The Circulation of Social Energy in Renaissance England.* Berkeley: University of California Press, 1988.

———. *Marvelous Possessions: The Wonder of the New World.* Chicago: University of Chicago Press, 1991.

Hahn, Thomas. "The Indian Tradition in Western Medieval Intellectual History." *Viator* 9 (1978): 213–34.

Hamel, Mary. "The 'Christening' of Sir Priamus in the Alliterative *Morte Arthure.*" *Viator* 13 (1982): 295–307.

———. "*The Siege of Jerusalem* as a Crusading Poem." In *Journeys to God: Pilgrimage and Crusade.* Ed. Barbara N. Sargent-Baur. Kalamazoo, MI: Medieval Institute Publications, 1992. 177–94.

Hamilton, George L. "Some Sources of the Seventh Book of Gower's *Confessio amantis.*" *Modern Philology* 9 (1912): 323–46.

———. "Studies in the Sources of Gower." *Journal of English and Germanic Philology* 26 (1927): 491–500.

Hanna, Ralph III. "Contextualizing *The Siege of Jerusalem.*" *YLS* 6 (1992): 109–21.

Hanna, Ralph III. "Alliterative Poetry." In *The Cambridge History of Medieval English Literature*. Ed. David Wallace. Cambridge: Cambridge University Press, 1999. 488–512.

———. "Langland's Ymaginatif: Images and the Limits of Poetry." In *Images, Idolatry and Iconoclasm in Late Medieval England: Textuality and the Visual Image*. Ed. Jeremy Dimmock, James Simpson, and Nicolette Zeeman. Oxford: Oxford University Press, 2002. 81–94.

Hartog, François. *The Mirror of Herotodus: The Representation of the Other in the Writing of History*. Trans. Janet Lloyd. Berkeley and Los Angeles: University of California Press, 1988.

Hartung, Albert E. and J. Burke Severs, eds. *A Manual of the Writings in Middle English, 1050–1500*. 9 vols. to date. New Haven: Connecticut Academy of Arts and Sciences, 1967–93.

Heffernan, Thomas J. "Orthodoxies Redux: The *Northern Homily Cycle* in the Vernon Manuscript and Its Textual Affiliations." In *Studies in the Vernon Manuscript*. Ed. Derek Pearsall. Cambridge: D.S. Brewer, 1990. 75–87.

Heng, Geraldine. *Empire of Magic: Medieval Romance and the Politics of Cultural Fantasy*. New York: Columbia University Press, 2003.

Higgins, Iain. *Writing East: The "Travels" of Sir John Mandeville*. Philadelphia: University of Pennsylvania Press, 1997.

Howard, Donald. "The World of *Mandeville's Travels*." *The Yearbook of English Studies* 1 (1971): 1–17.

———. *Writers and Pilgrims: Medieval Pilgrimage Narratives and Their Posterity*. Berkeley: University of California Press, 1980.

Humphries, K.W. "The Library of John Erghome and personal libraries of the fourteenth century in England." *Proceedings of the Leeds Philosophical and Literary Society*, Literary and Historical Section, 18 (1982): 106–23.

———, ed. *The Friars' Libraries*. CBMLC 1. London: The British Library, 1990.

Hunt, R.W. "A Manuscript Belonging to Robert Wivill, Bishop of Salisbury." *Bodleian Library Record* 7 (1962–67): 23–27.

Hussey, S.S., ed. *Piers Plowman: Critical Approaches*. London: Methuen, 1969.

Ito, Mayasoshi. "Gower's 'Diogenes and Alexander' and Its Philosophic-Literary Tradition." *Poetica* (Tokyo) 16 (1983): 66–77.

James, M.R. *The Ancient Libraries of Canterbury and Dover*. Cambridge: Cambridge University Press, 1903.

———. *A Descriptive Catalogue of the Manuscripts in the Library of Gonville and Caius College, Cambridge*. 2 vols. Cambridge: Cambridge University Press, 1907–08.

———. *A Descriptive Catalogue of the Manuscripts in the Library of Corpus Christi College, Cambridge*. 2 vols. Cambridge: Cambridge University Press, 1909–12.

Johnson, Lesley. "King Arthur at the Crossroads to Rome." In *Noble and Joyous Histories: English Romances, 1375–1650*. Ed. Eiléan Ní Cuilleanáin and J.D. Pheifer. Dublin: Irish Academic Press, 1993. 87–111.

Justice, Steven. *Writing and Rebellion*. Berkeley: University of California Press, 1994.

Kane, George. *Chaucer and Langland: Historical and Textual Approaches*. Berkeley and Los Angeles: University of California Press, 1989.

Keiser, George R. "The Theme of Justice in the Alliterative *Morte Arthure*." *Annuale Medievale* 16 (1975): 94–109.

————. "Lincoln Cathedral MS.91: Life and Milieu of the Scribe." *Studies in Bibliography* 32 (1979): 158–79.

————. "More Light on the Life and Milieu of Robert Thornton." *Studies in Bibliography* 36 (1983): 111–19.

Kelly, Henry Ansgar. "Jews and Saraens in Chaucer's England." *SAC* 27 (2005): 125–65.

Kerby-Fulton, Kathryn. "*Piers Plowman*." In *The Cambridge History of Medieval English Literature*. Ed. David Wallace. Cambridge: Cambridge University Press, 1999. 513–38.

Kerby-Fulton, Kathryn and Denise L. Despres. *Iconography and the Professional Reader: The Politics of Book Production in the Douce Piers Plowman*. Minneapolis: University of Minnesota Press, 1999.

Kimble, G.H.T. *Geography in the Middle Ages*. London: Methuen, 1938.

Knowles, M.D. *The Religious Orders in England II: The End of the Middle Ages*. Cambridge: Cambridge University Press, 1955.

————. "The Censured Opinions of Uthred of Boldon." *Proceedings of the British Academy* 37 (1951): 305–42.

Komroff, Manuel. *Contemporaries of Marco Polo*. New York: Boni & Liveright, 1928.

Kruger, Steven F. "The Spectral Jew." *NML* 2 (1998): 9–35.

Lampert, Lisa. *Gender and Jewish Difference from Paul to Shakespeare*. Philadelphia: University of Pennsylvania Press, 2004.

Langmuir, Gavin. *History, Religion, and Antisemitism*. Berkeley: University of California Press, 1990.

————. *Toward a Definition of Antisemitism*. Berkeley: University of California Press, 1990.

Lascelles, Mary. "Alexander and the Earthly Paradise in Mediaeval English Writings." *MAE* 5 (1936): 31–47, 69–104, 173–88.

Lawton, David. *Chaucer's Narrators*. Woodbridge, Suffolk: D.S. Brewer, 1985.

————. "Titus Goes Hunting and Hawking: The Poetics of Recreation and Revenge in *The Siege of Jerusalem*." In *Individuality and Achievement in Middle English Poetry*. Ed. O.S. Pickering. Cambridge: D.S. Brewer, 1997. 105–117.

————. "The Surveying Subject and the 'Whole World' of Belief: Three Case Studies." *NML* 4 (2001): 9–37.

Leff, Gordon. *Bradwardine and the Pelagians*. Cambridge: Cambridge University Press, 1957.

Lewis, C.S. *The Allegory of Love*. Oxford: Clarendon Press, 1936.

Lienard, E. "La Collatio Alexandri et Dindimi." *Revue Belge de Philologie et d'Histoire* 15 (1936): 819–38.

Little, Lester K. *Religious Poverty and the Profit Economy in Medieval Europe*. Ithaca: Cornell University Press, 1978.

Lomperis, Linda. "Medieval Travel Writing and the Question of Race." *JMEMS* 31 (2001): 147–64.

Loomis, R.S. "Alexander the Great's Celestial Journey." *The Burlington Magazine for Connoisseurs* 32 (1918): 136–140, 177–185.

Manly, J.M. Letter to the *Times Literary Supplement*, no. 1338 (Thursday, Sept. 22, 1927). 647.

Mann, Jill. "Chaucer and Atheism." *SAC* 17 (1995): 5–19.

Manzalaoui, M.A. "The Pseudo-Aristotelian *Kitab Sirr al-Asrar*: Facts and Problems." *Oriens* 23–24 (1974, for 1970–71): 148–257.

———. " 'Noght in the Registre of Venus': Gower's English Mirror for Princes." In *Medieval Studies for J.A.W. Bennett (Aetatis Suae LXX)*. Ed. P.L. Heyworth. Oxford: Clarendon Press, 1981. 159–83.

Marcett, Mildred Elizabeth. *Uhtred de Boldon, Friar William Jordan, and Piers Plowman*. New York: privately printed, 1938.

Margherita, Gayle. *Romance of Origins: Language and Sexual Difference in Middle English Literature*. Philadelphia: University of Pennsylvania Press, 1994.

May, David. "Dating the English Translation of Mandeville's Travels: The Papal Interpolation." *Notes and Queries* n.s. 34 (1987): 175–78.

McAlindon, T. "Hagiography into Art: A Study of *St. Erkenwald*." *SP* 67 (1970): 472–94.

Metlitzki, Dorothy. *The Matter of Araby in Medieval England*. New Haven: Yale University Press, 1977.

Meyer, Paul. *Alexandre le Grand dans la Littérature Française du Moyen Âge*. 2 vols. Paris: F. Vieweg, 1886.

Middleton, Anne. "The Audience and Public of 'Piers Plowman.' " In *Middle English Alliterative Poetry and Its Literary Background*. Ed. David Lawton. Cambridge: D.S. Brewer, 1982. 101–23.

———. "Narration and the Invention of Experience: Episodic Form in *Piers Plowman*." In *The Wisdom of Poetry: Essays in Early English Literature in Honor of Morton W. Bloomfield*. Ed. Larry D. Benson and Siegfried Wenzel. Kalamazoo, MI: Medieval Institute Publications, 1982. 91–122, 280–83.

———. "Introduction: The Critical Heritage." In *A Companion to Press Plowman*. Ed. John A. Alford Berkeley: University of California Press, 1988. 1–25.

———. "William Langland's 'Kynde Name': Authorial Signature and Social Identity in Late Fourteenth-Century England." In *Literary Practice and Social Change in Britain, 1380–1530*. Ed. Lee Patterson. Berkeley: University of California Press, 1990. 15–82.

———. "Langland's Lives: Reflections on Late-Medieval Religious and Literary Vocabulary." In *The Idea of Medieval Literature: New Essays on Chaucer and Medieval Culture in Honor of Donald R. Howard*. Ed. James M. Dean and Christian Zacher. Newark: University of Delaware Press, 1992. 227–42.

Millar, Bonnie. *The Siege of Jerusalem in Its Physical, Literary, and Historical Contexts*. Dublin: Four Courts Press, 2000.

Minnis, Alastair. "John Gower, *Sapiens* in Ethics and Politics." *MAE* 49 (1980): 207–29.

———. "Langland's Ymaginatif and late-medieval theories of the imagination." *Comparative Criticism* 3 (1981): 71–103.

———. *Chaucer and Pagan Antiquity*. Cambridge: D.S. Brewer, 1982.

———. "From Medieval to Renaissance? Chaucer's Position on Past Gentility." *Proceedings of the British Academy* LXXII (1986): 204–46.

———, Charlotte C. Morse, and Thorlac Turville-Petre, eds. *Essays on Ricardian Literature in Honour of J.A. Burrow*. Oxford: Clarendon Press, 1997.

———. "Looking for a Sign: The Quest for Nominalism in Chaucer and Langland." In *Essays on Ricardian Literature in Honour of J.A. Burrow*. Ed. Minnis, Morse, and Turville-Petre. Oxford: Clarendon Press, 1997. 142–76.

Mooney, Linne R. "A New Manuscript by the Hammond Scribe Discovered by Jeremy Griffiths." In *The English Medieval Book: Studies in Memory of Jeremy Griffiths.* Ed. A.S.G. Edwards, Vincent Gillespie, and Ralph Hanna III. London: The British Library, 2000. 113–23.

Moore, R.I. *The Formation of a Persecuting Society.* Oxford: Basil Blackwell, 1987.

Narin van Court, Elisa. "The Siege of Jerusalem and Augustinian Historians: Writing about Jews in Fourteenth-Century England." *Chaucer Review* 29 (1995): 227–48.

———. "The Hermeneutics of Supersession: The Revisions of the Jews from the B to the C text of *Piers Plowman.*" *YLS* 10 (1996): 43–88.

———. "Socially Marginal, Culturally Central: Representing Jews in Late Medieval English Literature." *Exemplaria* 12 (2000): 293–326.

Nicholson, Peter. *An Annotated Index to the Commentary on Gower's Confessio amantis.* Binghamton, NY: Medieval & Renaissance Texts and Studies, 1989.

Nicholson, Roger. "Haunted Itineraries: Reading *The Siege of Jerusalem.*" *Exemplaria* 14 (2002): 447–84.

Nissé, Ruth. " 'A Coroun Ful Riche': The Rule of History in *St. Erkenwald.*' *ELH: English Literary History* 65 (1998): 277–95.

Oberman, Heiko. "*Facientibus quod in se est Deus non denegat gratiam*: Robert Holcot, O.P, and the Beginning of Luther's Theology." *Harvard Theological Review* 55 (1962): 317–42.

Olschki, Leonard. *Marco Polo's Precursors.* Baltimore: Johns Hopkins Press, 1942.

Olsson, Kurt. *John Gower and the Structures of Conversion.* Cambridge: D.S. Brewer, 1992.

Otter, Monika. " 'New Werke': *St. Erkenwald*, St. Albans, and the medieval sense of the past." *JMRS* 24 (1994): 385–14.

Pantin, W.A. "Two Treatises of Uthred de Boldon on the Monastic Life." In *Studies in Medieval History Presented to Frederick Maurice Powicke.* Ed. R.W. Hunt, W.A. Pantin, and R.W. Southern. Oxford: Clarendon, 1948. 363–85.

———. "Some Medieval English Treatises on the Origins of Monasticism." In *Medieval Studies Presented to Rose Graham.* Ed. Veronica Raffer and A.J. Taylor. Oxford: Oxford University Press, 1950. 189–215.

Paris, Gaston. "La Légende de Trajan." *Mélanges de l'école des hautes études*, Fasc. 35e. Paris: École pratique des hautes études, 1878.

Patterson, Lee. *Negotiating the Past: The Historical Understanding of Medieval Literature.* Madison, WI: University of Wisconsin Press, 1987.

———. "Introduction: Critical Historicism and Medieval Studies." In *Literary Practice and Social Change in Britain, 1380–1530.* Ed. Lee Patterson. Berkeley and Los Angeles: University of California Press, 1990. 1–14.

———. "On the Margin: Postmodernism, Ironic History, and Medieval Studies." *Speculum* 65 (1990): 87–108.

———. *Chaucer and the Subject of History.* Madison, WI: University of Wisconsin Press, 1991.

———. " 'The Living Witness of Our Redemption': Martyrdom and Imitation in Chaucer's *Prioresse's Tale.*" *JMEMS* 31:3 (2001): 507–60.

Pearsall, Derek, ed. *Studies in the Vernon Manuscript.* Cambridge: D.S. Brewer, 1990.

———. *The Life of Geoffrey Chaucer.* Oxford: Blackwell, 1992.

Peck, Russell A. *Kingship and Common Profit in Gower's Confessio amantis.* Carbondale, IL: Southern Illinois University Press, 1978.

Pfister, F. "Auf den Spuren Alexanders des Grossen in der älteren englishen Literatur." *Germanisch-Romanische Monatsshchrift* 16 (1928): 81–86.

Porter, Elizabeth. "Gower's Ethical Microcosm and Political Macrocosm." In *Gower's Confessio amantis: Responses and Reassessments.* Ed. A.J. Minnis. Cambridge: D.S. Brewer, 1983. 153–60.

Pratt, Robert A. "Some Latin Sources of the Nonnes Preest on Dreams." *Speculum* 52 (1977): 538–70.

Rhodes, Jim. *Poetry Does Theology: Chaucer, Grosseteste, and the Pearl-Poet.* Notre Dame, IN: Notre Dame University Press, 2001.

Rigg, A.G. "John of Bridlington's Prophecy: A New Look." *Speculum* 63 (1988): 596–613.

Ritze-Rutherford, Jean. "Formulaic Macrostructure: The Theme of Battle." In *The Alliterative Morte Arthure: A Reassessment of the Poem.* Ed. Karl Heinz Göller. Cambridge: D.S. Brewer, 1981. 83–95.

Rizzo, Gino. "Dante and the Virtuous Pagans." In *A Dante Symposium, in Commemoration of the 700th Anniversary of the Poet's Birth (1265–1965).* Ed. William De Sua and Gino Rizzo. University of North Carolina Studies in the Romance Languages and Literatures 58. Chapel Hill, NC: University of North Carolina Press, 1965. 115–40.

Rooney, Ellen. "Form and Contentment." *Modern Language Quarterly* 61 (2000): 17–40.

Ross, D.J.A. *Alexander Historiatus: A Guide to Medieval Illustrated Alexander Literature.* Warburg Institute Surveys 1. London: Warburg Institute, 1963.

———. "*Parva Recapitulatio*: An English Collection of Texts relating to Alexander the Great." *Classica et Medievalia* 33 (1981–82): 191–203.

Rossabi, Morris. *Khubilai Khan: His Life and Times.* Berkeley: University of California Press, 1988.

Ruddy, David Wilmot. "Scribes, Printers and Vernacular Authority: A Study in the Late-Medieval and Early-Modern Reception of *Mandeville's Travels.*" Ph.D. Diss., University of Michigan, 1995.

Russell, G.H. "The Salvation of the Heathen: The Exploration of a Theme in *Piers Plowman.*" *Journal of the Warburg and Courtald Institutes* 29 (1966): 101–116.

Said, Edward. *Orientalism.* New York: Random House, 1994.

Scanlon, Larry. *Narrative, Authority and Power: The Medieval Exemplum and the Chaucerian Tradition.* Cambridge: Cambridge University Press, 1994.

Scase, Wendy. *Piers Plowman and the New Anticlericalism.* Cambridge: Cambridge University Press, 1989.

Schmidt, Victor M. *A Legend and Its Image: The Aerial Flight of Alexander the Great in Medieval Art.* Trans. Xandra Bardet. Medievalia Groningana, fasc. 17 Groningen: E. Forsten, 1995.

Scott, Kathleen L. *Later Gothic Manuscripts, 1390–1490.* 2 vols. London: H. Miller, 1996.

Seymour, M.C. "Mandeville and Marco Polo: A Stanzaic Fragment." *AUMLA: Journal of the Australasian Universities Language and Literature Association* 21 (1964): 39–52.

———. *Sir John Mandeville.* Authors of the Middle Ages 1. Aldershot: Variorum, 1994.

Sharpe, R., J.P. Carley, R.M. Thomson, and A.G. Watson. *English Benedictine Libraries: The Shorter Catalogues*. CBMLC 4. London: The British Library, 1995.

Simon, M. "Alexandre le Grand, juif et chretien." *Revue d'Histoire et Philosophie Religieuses* XXI (1941): 177–91.

Simpson, James. *Piers Plowman: An Introduction to the B-Text*. London: Longmans, 1990.

———. *Sciences and the Self in Medieval Poetry: Alan of Lille's Anticlaudianus and John Gower's Confessio amantis*. Cambridge: Cambridge University Press, 1995.

Smith, D. Vance. "Crypt and Decryption: *Erkenwald* Terminable and Interminable." *NML* 5 (2002): 59–85.

Spearing, A.C. "A Ricardian 'I': The Narrator of 'Troilus and Criseyde." In *Essays in Ricardian Literature*. Ed. Minnis, Morse, and Turville-Petre. 1–22.

Steiner, Emily. *Documentary Culture and the Making of Medieval English Literature*. Cambridge: Cambridge University Press, 2003.

Taylor, E.G.R. "The Cosmographical Ideas of Mandeville's Day." In *Mandeville's Travels: Texts and Translations*. Ed. Letts. I.li–lix.

"Televangelist Larry Lea's Followers Speak." *San Francisco Chronicle/Examiner*, Sunday, Nov. 18, 1990.

Thompson, David. "Dante's Virtuous Pagans." *Dante Studies* 96 (1973): 145–62.

Thompson, John J. *Robert Thornton and the London Thornton Manuscript: British Library MS Additional 31042*. Cambridge: D.S. Brewer, 1987.

Tomasch, Sylvia. "Judecca, Dante's Satan, and the Dis-placed Jew." In *Text and Territory: Geographical Imagination in the European Middle Ages*. Ed. Sylvia Tomasch and Sealy Gilles. Philadelphia: University of Pennsylvania Press, 1988. 247–67.

———. "Postcolonial Chaucer and the Virtual Jew." In *The Postcolonial Middle Ages*. Ed. Jeffrey Jerome Cohen. New York: St. Martin's Press, 2000. 243–60.

Tomasch, Sylvia and Sealy Gilles, eds. *Text and Territory: Geographical Imagination in the European Middle Ages*. Philadelphia: University of Pennsylvania Press, 1998.

Trigg, Stephanie. *Congenial Souls: Reading Chaucer from Medieval to Postmodern*. Minneapolis: University of Minnesota Press, 2002.

Turville-Petre, Thorlac. *The Alliterative Revival*. Cambridge: D.S. Brewer, 1977.

Vickers, Nancy. "Seeing Is Believing: Gregory, Trajan, and Dante's Art." *Dante Studies* 101 (1983): 67–85.

Vitto, Cindy L. "The Virtuous Pagan in Middle English Literature." *Transactions of the American Philosophical Society* 79 (1989).

Voorbij, J.B. "The *Speculum Historiale*: some aspects of its genesis and manuscript tradition." In *Vincent of Beauvais and Alexander the Great*. Ed. Aerts et al. 11–56.

Wallace, David, ed. *The Cambridge History of Medieval English Literature*. Cambridge: Cambridge University Press, 1999.

Waller, J.G. "The Lords of Cobham, Their Monuments, and the Church." *Archaeologica Cantiana* XI (1877): 49–112.

Ward, A.W. and A.P. Waller, eds. *The Cambridge History of English Literature*. Vol. 2: *The End of the Middle Ages*. Cambridge: Cambridge University Press, 1908.

Ward, H.L.D. *Catalogue of Romances in the Department of Manuscripts in the British Museum*. 3 vols. London: Trustees of the British Museum, 1883–1910.

Warm, Robert. "Arthur and the Giant of Mont St. Michel: The Politics of Empire Building in the Later Middle Ages." *Nottingham Medieval Studies* 41 (1997): 57–71.

Warner, George and Julius P. Gilson. *Catalogue of the Western Manuscripts in the Old Royal and King's Collections.* 4 vols. London: Trustees of the British Museum, 1921.

Watson, Andrew. *Medieval Libraries of Great Britain. . .: A Supplement to the Second Edition.* London: Offices of the Royal Historical Society, 1987.

Watson, Nicholas. "Censorship and Cultural Change in Late-Medieval England: Vernacular Theology, the Oxford Translation Debate, and Arundel's Constitutions of 1409." *Speculum* 70 (1995): 822–64.

———. "Visions of Inclusion: Universal Salvation and Vernacular Theology in Pre-Reformation England." *JMEMS* 27 (1997): 145–87.

———. "Desire for the Past." *SAC* 21 (1999): 59–97.

Watt, Diane. *Amoral Gower: Language, Sex and Politics.* Minneapolis: University of Minnesota Press, 2003.

Weiss, R. *Humanism in England During the Fifteenth Century.* Oxford: Blackwell, 1957.

Wells, John Edwin. *A Manual of Writings in Middle English, 1050–1400.* New Haven: Connecticut Academy of Arts and Sciences, 1916.

Wenzel, Siegfried. "Langland's Troianus." *YLS* 10 (1996): 181–85.

Westrem, Scott. "Against Gog and Magog." In *Text and Territory: Geographical Imagination in the European Middle Ages.* Ed. Sylvia Tomasch and Sealy Gilles. Philadelphia: University of Pennsylvania Press, 1988. 54–75.

Wetherbee, Winthrop. *Chaucer and the Poets: An Essay on Troilus and Criseyde.* Ithaca, NY: Cornell University Press, 1984.

———. "John Gower." In *The Cambridge History of Medieval English Literature.* Ed. Wallace. 589–609.

Whatley, Gordon. "The Middle English *St. Erkenwald* and Its Liturgical Context." *Mediaevalia* 8 (1982): 278–306.

———. "*Piers Plowman* B 12.277–94: Notes on Language, Text, Theology." *Modern Philology* 82 (1984): 1–12.

———. "The Uses of Hagiography: The Legend of Pope Gregory and the Emperor Trajan in the Middle Ages." *Viator* 15 (1984): 25–63.

———. "Heathens and Saints: *St. Erkenwald* in Its Legendary Context." *Speculum* 61 (1986): 330–63.

Wheeler, Bonnie. "Dante, Chaucer, and the Ending of *Troilus and Criseyde.*" *Philological Quarterly* 61 (1982): 105–23.

White, Hayden. *The Content of the Form: Narrative Discourse and Historical Representation.* Baltimore: Johns Hopkins University Press, 1987.

Windeatt, Barry. *Oxford Guides to Chaucer: Troilus and Criseyde.* Oxford: Oxford University Press, 1992.

Wittig, Joseph. "'Piers Plowman' B, Passus IX-XII: Elements in the Design of the Inward Journey." *Traditio* 28 (1972): 211–80.

Wogan-Browne, Jocelyn, Nicholas Watson, Andrew Taylor, and Ruth Evans, eds. *The Idea of the Vernacular: An Anthology of Middle English Literary Theory, 1280–1520.* University Park, PA: Pennsylvania State University Press, 1999.

Woolf, Rosemary. "The Tearing of the Pardon." In *Piers Plowman: Critical Approaches.* Ed. S.S. Hussey. London Methuen, 1969. 50–75.

Yeager, R.F. "*Pax Poetica*: On the Pacifism of Chaucer and Gower." *SAC* 9 (1987): 97–121.

———. "English, Latin, and the Text as 'Other': The Page as Sign in the Work of John Gower." *Text* 3 (1987): 251–67.

———. *John Gower's Poetic: The Search for a New Arion*. Cambridge: D.S. Brewer, 1990.

Yeager, Suzanne M. "*The Siege of Jerusalem* and Biblical Exegesis: Writing about Romans in Fourteenth-Century England." *ChR* 39.1 (2004): 70–102.

Young, John and P. Henderson Aitken. *A Catalogue of the Manuscripts in the Library of the Hunterian Museum in the University of Glasgow*. Glasgow: J. Maclehose and Sons, 1908.

Zacher, Christian K. *Curiosity and Pilgrimage*. Baltimore: Johns Hopkins University Press, 1976.

Ziolkowski, Jan. "A Narrative Structure in the Alliterative Morte Arthure, 1–1221 and 3150–4346." *ChR* 22 (1988): 234–45.

INDEX